DDG-51 Engineering Training

How Simulators Can Help

T0195287

Roland J. Yardley, James G. Kallimani,
Laurence Smallman, Clifford A. Grammich

Prepared for the United States Navy
Approved for public release; distribution unlimited

 NATIONAL DEFENSE RESEARCH INSTITUTE

The research described in this report was prepared for the United States Navy. The research was conducted in the RAND National Defense Research Institute, a federally funded research and development center sponsored by the Office of the Secretary of Defense, the Joint Staff, the Unified Combatant Commands, the Department of the Navy, the Marine Corps, the defense agencies, and the defense Intelligence Community under Contract W74V8H-06-C-0002.

Library of Congress Cataloging-in-Publication Data is available for this publication.

978-0-8330-4729-8

Published 2009 by the RAND Corporation
1776 Main Street, P.O. Box 2138, Santa Monica, CA 90407-2138
1200 South Hayes Street, Arlington, VA 22202-5050
4570 Fifth Avenue, Suite 600, Pittsburgh, PA 15213-2665
RAND URL: http://www.rand.org/
To order RAND documents or to obtain additional information, contact
Distribution Services: Telephone: (310) 451-7002;
Fax: (310) 451-6915; Email: order@rand.org

Preface

U.S. Navy surface combatant crews undergo extensive and rigorous training to operate their ships. The training needs are challenging for crews on DDG-51–class destroyers, the most numerous ships among the surface ship combatants, and specifically for engineers on these craft. Much of this training is done underway, but most can be done in port or on simulators at considerable savings.

The Director, Assessment Division (OPNAV N81) therefore asked the RAND Corporation to examine the training requirements for DDG-51 engineering watchstanders, specifically, how available engineering simulation technology might be adapted for use by DDG-51–class ship crews and what policies and resources could help increase the use of simulators for engineering training. This monograph reports our findings. It should interest those concerned with the training and readiness of Navy surface combatants, including members of the Fleet Forces Command, the Type Commander, and the broader defense operational planning and budgeting community.

This research was sponsored by OPNAV N81 and conducted within the Acquisition and Technology Policy Center of the RAND National Defense Research Institute, a federally funded research and development center sponsored by the Office of the Secretary of Defense, the Joint Staff, the Unified Combat Commands, the Department of the Navy, the Marine Corps, the defense agencies, and the defense Intelligence Community.

For more information about this work, contact Roland Yardley. He can be reached by email at yardley@rand.org or by phone at

703-413-1100, extension 5578. For information on RAND's Acquisition and Technology Policy Center, contact the Director, Philip Antón. He can be reached by email at atpc-director@rand.org; by phone at 310-393-0411, extension 7798; or by mail at the RAND Corporation, 1776 Main Street, P. O. Box 2138, Santa Monica, California 90407-2138. More information about RAND is available at www.rand.org.

Contents

APPENDIXES

Figures

Tables

Summary

U.S. Navy surface combatant ship crews require extensive and rigorous training. The training demands are many on ships of the DDG-51 class. Among DDG-51 crew members, some of the most rigorous training is required for ship engineers responsible for maintaining, operating, and repairing main propulsion and auxiliary equipment.

The basics of engineering training consist in developing watchstander proficiency in two different skill sets. The first required skill set is the ability to respond to engineering casualties. This training consists of executing a series of engineering drills, coordinated by the ship's Engineering Training Team (ETT), during which the watchstanders must respond to the symptoms of the casualty, take the correct controlling and immediate actions from memory, and then restore the plant to its normal operating configuration. This training is time-consuming and repetitive. All members of the watch section must function effectively as individuals and as a team. It takes repeated exposure to understand the casualty and to learn and memorize the actions needed to correct, control, and recover the engineering plant. The use of simulators has great value for this skill set.

The second required skill set is the ability to conduct routine plant operations or engineering evolutions[1] (i.e., starting and stopping various pumps, motors, and engines and aligning systems for use). During evolutions, the watchstander is required to have and refer to a written procedure while conducting the event. Evolutions can also be practiced

---

[1] *Evolutions* are events performed during the normal operation of the engineering plant.

on available engineering simulators—both on the desktop and on console trainers. Because evolutions are essentially "open book" tests, evolution proficiency is easier to achieve than proficiency in responding to engineering casualties.

Ships progress through a sequence of training events before being assigned to operational missions. Ships begin first with unit-level training (ULT), during which the ship's crews are assessed, trained and certified in the missions that the ship was designed to perform. After ships complete ULT, they are ready for tasking (RFT) as individual units and progress to intermediate and advanced training, where they train and operate with other ships and units. Upon completion of advanced training, ships are ready for deployed operations. Ships must sustain the training readiness achieved during ULT throughout the operational cycle. Our research focuses on the engineering training requirements and proficiency of engineering watchstanders.

Previous RAND research found that much of the ULT is conducted underway but that a great deal of it could be done in port.[2] Although underway training is arguably the best method for training a crew, it is expensive. While underway, ships burn large quantities of fuel and incur equipment wear and tear that may increase maintenance demands. The average DDG burns a minimum of 500 barrels (21,000 gallons) of fuel per 24-hour underway day. At an optimistic price of $50 per barrel, one can see that fuel costs of $25,000 per day per DDG are easily achieved.[3] Time constraints and other factors also limit how much underway training a crew can do. By contrast, simulated training could expand training opportunities. The use of a shore-based engineering simulator console could improve watchstanders' proficiency throughout the length of their tour on the ship, reduce the necessary ULT underway training days required for them to achieve satisfactory proficiency and thus saving fuel, reduce equipment

---

[2]   Roland J. Yardley, Harry J. Thie, Christopher Paul, Jerry M. Sollinger, and Alisa Rhee, *An Examination of Options to Reduce Underway Training Days Through the Use of Simulation,* Santa Monica, Calif.: RAND Corporation, MG-765-NAVY, 2008.

[3]   When the fully burdened cost of fuel to the Navy is considered, the fuel costs per underway day dramatically increase. Fully burdened fuel costs include costs to transport fuel to the fleet.

wear and tear, and potentially result in the ship being RFT earlier in the training cycle.

Recognizing these issues and the potential of simulated training to supplement underway training, the Director, Assessment Division (OPNAV N81) asked RAND to examine the training requirements for DDG-51 engineering watchstanders—specifically, how available simulation technology might be adapted for use by DDG-51–class ship engineers and what policies and resources could help increase the use of simulated training. Accordingly, this monograph reviews the ways that simulators can boost training proficiency as well as the changes needed to support their widespread adoption.

Current Training Challenges and How Simulators Could Meet Them

Subsequent to a maintenance period and before undergoing advanced training and deploying, ship crews—including ship engineers, the focus of this study—go through ULT to assess their readiness and mastery of drills and evolutions they are expected to handle in the conduct of routine plant operations. Typically, most ship engineering teams do not start at the required level of mastery during the initial assessment of these drills and must begin a period of mobility–engineering (MOB-E) training. Most ships entering MOB-E training complete it within three to four weeks, but a few in recent years have taken as many as six weeks or more to complete this training. Ships typically have three teams of engineering watchstanders, with the third team consisting of members of the Engineering Training Team (ETT) who are responsible for evaluating the other two.

Our research has shown that ship crews perform a majority of their training underway but that many training exercises could be done in port. This monograph discusses how simulator use could improve engineering watchstanders' proficiency before ships go to sea, so that time at sea could be used to fine-tune the training. Furthermore, given constraints on underway training—including other tasks that a ship must accomplish at sea as well as the resources needed to be at sea—

there is a limit to the drills a ship can practice at sea. In addition, the requirement that a third team train and evaluate the other two leaves little opportunity for it to conduct its own drills.

Simulation technology is currently available in three forms for a DDG-51 destroyer: a desktop trainer, a full mission console, and an embedded training capability (onboard and integrated into the ship's engineering consoles). The desktop trainer and full mission console use the same software, developed exclusively for the DDG-51. The full mission consoles include exact duplicates of the consoles onboard DDG-51 ships and provide training for watchstanders in normal startup, operations, shutdown, and casualty control procedures of a DDG-51 engineering plant. The desktop trainer can be operated either individually—to train operators on the seven DDG-51 engineering consoles to align, start, operate, or stop equipment—or in a local area network to provide watch team training. The embedded training capability is installed only on the newest ships of the class (DDG-96 and above), but plans are being made to backfit the embedded trainer on the older ships (DDG-51–DDG-95). The embedded training capability allows operators to train onboard their ship, on their own ship's consoles. The consoles are put in a training mode and an instructor inserts casualties via a laptop connection to the console, to evaluate the watchstander's responses.

Simulators, both onboard and ashore, can help increase an engineering watchstander's proficiency by allowing the ship's company to practice more drills and to practice each drill more frequently with fewer time constraints and less manpower.

Current simulators allow practice on 35 of 40 casualty control drills. Simulators on shore at Fleet Concentration Areas (FCAs) would also provide more-accessible training opportunities. The only current DDG-51–class engineering simulators ashore are those at the Surface Warfare Officers School in Newport, Rhode Island, and therefore they are not used by the enlisted personnel most likely to stand engineering watches. The Navy is currently backfitting embedded simulators on DDG-51–class ships but, at the current rate, will not complete this process until 2025.

Benefits and Drawbacks of Simulator Training

In addition to providing more training opportunities, simulators offer many unique advantages that do, in fact, provide for more-effective training. Their replay capability allows engineers to pinpoint the exact point at which a drill failed. The "freeze" capability allows an instructor to stop a drill when needed and provide instruction. They allow engineers to practice multiple or cascading casualties more easily than underway training would allow. They allow more repetitions of a single drill in much less time than is required underway, improving the pace of learning. More-qualified engineers can maintain their skills on simulators; less-experienced ones can gain experience and proficiency prior to underway training. Training via a simulator is a safer way to train. The impact of a trainee's incorrect actions or inactions will not harm equipment, personnel, or the ship.

We compared and contrasted the conduct of engineering casualty control drill training as it is done on the ship at sea, on the ship pierside, or in a shore-based simulator. Table S.1 compares these methods using a stoplight format—green being good or best, yellow being neutral, and red being poor or least attractive—by variables such as cost, training constraints, and cohesiveness.

We do not weight these variables, but we understand that some, e.g., cost, are more important than others. The table shows that shore-based simulators offer many training options that compare favorably with training done either onboard at sea or pierside in port. Advantages include lower cost, better cueing of watchstanders, better trainee feedback, reduced energy use, and increased training safety. These advantages do not suggest that a shore-based simulator is the single best option for conducting drills. Rather, such simulators could be part of a balanced approach to improving training.

Although simulators can offer some advantages over underway training in cost, scheduling of training, and trainee feedback, their use can be affected by several factors. Among the most important factors that support increased use of simulation are the close proximity to ship's engineers, high fidelity of simulator exercises compared with actual operating conditions, flexible times for use, and an adequate

Table S.1
Factors to Consider in Using Shore-Based Simulators or Shipboard Equipment for Training

Factor/Location of Training	At Sea	Pierside	Shore-Based Simulator
Cost	High fuel costs plus wear and tear	Lower cost, but wear and tear	Lower cost, no wear and tear
Operate own ship's equipment	All engineering equipment can be operated	Some can be operated, but not all	Ship's equipment not operated
Cueing of watchstanders	Some cueing by training team on drill imposition	Some cueing by training teams on drill imposition	No cueing
Number of ECC drills than can be done	All 40	32 of 40	35 of 40
Time available by crew for training	Dedicated crew underway, but underway time is decreasing	Maintenance demands in port are high. CCS is hub of activity in port— conflicts will arise	No conflicts, but competes with other unit's training needs
Training constraints	ECC drills normally done underway on a not-to-interfere basis with other training needs and/or impact bridge operations or ship's ability to navigate	Some conflicts with in-port maintenance demands and other ship events	Trainees must leave ship for training. Must trade off what they would be doing if they stayed on board, and what doesn't get done
Who gets trained	2 of 3 Engineering Watch Teams composed of CCS and in-space watchstanders	2 of 3 Engineering Watch Teams composed of CCS and in-space watchstanders	3 of 3 CCS watchstanders but not in-space watchstanders
Personnel involved in training	ETT and all watchstanders	ETT and all watchstanders	CCS personnel
Impact on nonengineering watchstanders	Electrical load limitations for combat systems, navigation and bridge equipment	Small impact	None
Usefulness to utilize for varying skill levels	Good for experienced and inexperienced personnel, but expensive and potentially hazardous if incorrect actions taken	Good for experienced and inexperienced personnel and less expensive; potentially hazardous if incorrect actions taken	Good for experienced and inexperienced personnel and least expensive over time; good for continuation training
Impact of watchstander errors	CCS personnel and in-space watchstanders – potential for being costly and dangerous	CCS personnel and in-space watchstanders— potential for being costly and dangerous	Trains CCS personnel only— no hazard to personnel or equipment

Table S.1—Continued

Factor/Location of Training	At Sea	Pierside	Shore-Based Simulator
Feedback mechanism to trainees	In-space watchstanders stopped for safety violations. CCS watchstanders will perform immediate and controlling actions—graded as effective or ineffective based on observation and written comments about their actions	In-space watchstanders stopped for safety violations. CCS watchstanders will perform immediate and controlling actions—graded as effective or ineffective based on observation and written comments about their actions	Program can be "frozen" to provide instruction to watchstanders. Printout of time and sequence of actions offer ability to trace actions and timeline and provide objective feedback
Time it takes to conduct training	Longer. Must be approved by commanding officer and deconflicted with other training events onboard	Long. Deconfliction is required with ongoing maintenance and other shipside training	Short. Provides list of drills and runs training events. Events may be repeated to ensure proficiency
Maintenance of Engineering Training Team (ETT) Casualty Control Proficiency	Proficiency of ETT is unknown and untested	Proficiency of ETT is unknown and untested	Good. ETT members receive proficiency training as well as 1st and 2nd watch teams; ECC drill proficiency can be maintained in a shore-based simulator
Engineering watchstander's cohesion	Good. All are trained and communicate together	Good. All are trained and communicate together	Good for CCS watchstanders only
Physiological—heat, sound, sight, smell, ship movement, stresses	The real thing	Fewer stresses in port	Simulated environment
Realism of drill imposition	Some impositions different from an actual casualty, e.g., grease pencil used to indicate a high tank level	Simulations and deviations exist	Casualties alarm and occur to CCS watchstanders as they would underway
Effectiveness standard	Underway demonstration standard is 50%	Onboard demonstration standard is 50%	Can be trained to a higher effectiveness standard
Energy savings/carbon footprint	High energy use	Reduced energy use	Little energy use
Safety	Proficiency gained on operating equipment	Proficiency gained on operating equipment	Safe. Can train and gain proficiency before getting underway

return on investment offsetting the costs of equivalent underway training. By contrast, a simulator at a removed distance that offers low or poor fidelity to actual operations and a limited range of exercises at high cost will hinder or limit simulated training. Overall, our research indicates that simulators should be used as a training alternative when they can sustain readiness, enhance a capability, save resources, or reduce risk.

Resources Needed to Increase Use of Simulators

We recommend installation of engineering full mission consoles at FCAs, such as Norfolk and San Diego. A DDG-51 engineering console trainer, such as the one used at the Surface Warfare Officers School, costs $1.6 million to procure and $300,000 to install. Sustainment costs include an operator, a technician, and updating the software as needed. The software for the desktop trainer can be loaded onto ship computers today at negligible cost to the Navy. Such software can be used to practice many console operations.

The payoff for installing full mission consoles at FCAs depends on the cost of resources and the number of underway days of training the console saves. For example, DDG-51–class ships burn a minimum of 500 barrels (21,000 gal) of fuel per day. If fuel were to cost $50 per barrel, then it would cost approximately $25,000 per 24-hour steaming day per DDG-51. A console that saves a total of 120 steaming days of training over the course of a decade would save $3 million in fuel costs alone ($25,000 × 120 = $3 million). These savings would pay for the simulator's acquisition and sustainment costs. In addition, there would be reductions in necessary ship maintenance, repairs, food costs, port costs, etc. We estimate that approximately 50 Norfolk-based ships and 39 San Diego–based ships (without an embedded training capability) will undergo ULT from fiscal year (FY) 2009 to FY 2018. Even if fuel were just $40 per barrel over the next decade, an engineering simulator in Norfolk would pay for itself if it saved only about three days per ship of underway training over a ten-year period, while one in San Diego would pay for itself if it saved only about four days per ship.

We recognize that, when a DDG is underway for MOB-E training, engineering is not the only training the ship conducts. However, our discussions with Afloat Training Group (ATG) representatives on both coasts indicate that MOB-E training, when it is in a ship's Plan of the Week for an underway week, is normally the preponderance of the effort. MOB-E training is a major driver for underway training, and simulators will reduce that requirement.

Policy changes could further encourage use of simulators for training. To fully realize the benefits of the engineering simulator, its use should be a mandatory part of the training process. To increase engineering training through simulators, the Navy should, among other steps, establish console trainers at FCAs, using them for training during extended maintenance periods and repetitive training requirements; use desktop trainers as lead-in trainers for advanced console operations; load the engineering training software onboard ships or ashore and at homeports without console trainers; and perform alignment, starting/stopping, and master light-off checklist plant operations on the desktop trainers.

The DDG-51 community could consider utilizing simulators to qualify/requalify senior enlisted personnel who are reporting back to sea duty from a shore-duty assignment. A refresher course would allow personnel to arrive at their new commands ready for qualification. It would free up senior engineering talent to focus more on monitoring material condition and training and mentoring subordinates.

Acknowledgments

We would like to thank the staff of OPNAV N81 for their support, especially Mr. Carlton Hill, CAPT Catherine Osman, CAPT Eric Kaniut, and CAPT James Brown. We are grateful for the support provided by Mr. Stephen Williams, who assisted our efforts throughout the project. We appreciate the guidance and insights provided by RADM Brian C. Prindle, RADM Daniel Cloyd, and Mr. Trip Barber.

Our efforts were supported by the U.S. Navy training commands responsible for DDG-51 engineering training. We appreciate the time and advice provided to us by CAPT Mark Hoyle, CDR Dave Morris, and LCDR Chris Ledlow of ATG, Atlantic; and by CAPT Mike Taylor and CDR Tom Shaw of ATG, Pacific. We appreciate the insights and guidance provided by RADM Victor Guillory, OPNAV N86, RADM Philip Cullom, OPNAV N43, and Commodore Perry Bingham, DDG Class Squadron Commander. We also appreciate the thoughts provided by CAPT Ken Krogman of Commander, Naval Surface Forces, Atlantic. Mr. Dan Rodgers of Commander, Naval Surface Forces, Pacific, also assisted the research by addressing issues related to scheduling surface combatant underway training.

We are indebted to many engineering specialists outside the U.S. Navy for their willingness to assist and generosity with their time. We list here only our main contacts but wish to extend our thanks to their colleagues who helped us as well: Mr. Chuck Eser, Manager Academic Affairs, Calhoon MEBA Engineering School; Mr. Gerry Miante, Chief Engineer, Assistant Commandant for Prevention (CG-5221) and LCDR Pete McCaffrey, Training & Performance Management

Commandant (CG-132) of the U.S. Coast Guard; LCDR Paul Busatta CD, Assistant Naval Attaché (Engineering) of the Canadian Navy; CDR Paul Marshall RN, Assistant Naval Attaché of the British Royal Navy; CDR Mark Worsfold, RNZN, Assistant Naval Attaché and Mr. Henry Cameron, Fleet Personnel and Training Organisation of the Royal New Zealand Navy.

We appreciate the time and effort of the Surface Warfare Officers School (SWOS) Executive Director, Mr. George Ponsolle, and his engineering staff, who arranged our tour of SWOS facilities and discussed the school's training and engineering simulator capabilities. Mr. Finn Kilsgaard of Naval Air Systems Command Training Systems Division Orlando, was instrumental in our research, providing us with the details of the capabilities, limitations, and costs of the DDG-51 engineering simulators at SWOS.

We appreciate the insights provided throughout the project by RAND colleague Harry Thie. We are grateful to CAPT Andy Diefenbach, USN (Ret.) and RAND colleague LCDR Dan Cobian for their extensive and thoughtful suggestions on an early draft of the monograph.

We also thank Miriam Polon for editing the manuscript, Matthew Byrd for coordinating the document's production, Erin-Elizabeth Johnson for typesetting the book, and Carol Earnest for her work on the figures. We acknowledge and appreciate the administrative support provided by Christine Galione.

The views expressed herein are our own and do not necessarily represent the policy of the Department of the Navy.

Abbreviations

AEM	auxiliary equipment monitor
ASM	auxiliary systems monitor
ATG	Afloat Training Group
ATGLANT	Afloat Training Group, Atlantic
ATGPAC	Afloat Training Group, Pacific
AUX	auxiliary
BCU	bridge control unit
BS&W	bottom sediment and water
CCS	central control station
CNE	Center for Naval Engineering
CO	commanding officer
CPO	Chief Petty Officer
CPP	controllable pitch propeller
CPS	central processing system
CRP	controllable reversible pitch
CY	calendar year
DCC	damage control console

DDGRON	destroyer class squadron
DESRON	destroyer squadron
ECC	engineering casualty control
EDORM	engineering department organization and regulations manual
EEBD	emergency escape breathing device
EOC	Engineering Operational Certification
EOCC	Engineering Operational Casualty Control
EOOW	engineering officer of the watch
EOP	Engineering Operational Procedures
EOSS	Engineering Operations Sequencing System
EOT	engine order telegraph
EPCC	electric plant control console
ERO	engine room operator
ETT	Engineering Training Team
FCA	Fleet Concentration Area
F/O	fuel oil
FY	fiscal year
GTG	gas turbine generator
GTM	gas turbine module
GTPPT	Gas Turbine Propulsion Plant Trainer
HPAC	high-pressure air compressor
ISIC	immediate superior in command
L/O	lube oil

LPAC	low-pressure air compressor
LPAD	low-pressure air dehydrator
MCS	Machinery Control System
MER	main engine room
MLOC	Master Light-Off Checklist
MOB-D	mobility–damage control
MOB-E	mobility–engineering
MRG	main reduction gear
NAWC TSD	Naval Air Warfare Command, Training Systems Division
OBT	onboard trainer
PACC	propulsion and auxiliary control console
PLA	power level actuator
PLC	Programmable Logic Controller
PLCC	propulsion local control console
PMS	Planned Maintenance System
PO	Petty Officer
PQS	Personnel Qualification Standards
PSM	Propulsion Systems Monitor
RFT	ready for tasking
RIMSS	Redundant Independent Mechanical Start System
S&S	sounding and security
SCC	ship control console

SCU	shaft control unit
SSGTG	ships' service gas turbine generator
STE	Synthetic Training Environment
S/W	sea water
SWBD OP	Switchboard Operator
SWOS	Surface Warfare Officers School
TSTA	Tailored Ship Training Availability
ULT	unit-level training
ULTRA-E	ULT Readiness Assessment–Engineering
WDCM	washdown countermeasures

CHAPTER ONE

Introduction

U.S. Navy surface combatant ship crews require extensive and rigorous training. These training needs are demanding for crew members on destroyers, or ships of the DDG-51 class, the most numerous craft among the Navy's surface combatant fleet. Especially rigorous training is required for ship engineers responsible for operating, maintaining, and repairing main propulsion and auxiliary equipment to keep the ship ready to go to sea. Engineers must be trained and qualified to operate equipment, proficient in standing engineering watches, and able to perform engineering casualty control procedures.

Earlier RAND research found that much of this training is conducted underway.[1] Although underway training is arguably the best method for training a crew, it is expensive and becoming ever more costly. While underway for training, a DDG-51 burns more than 500 barrels of fuel daily and uses other consumables, such as lubricating oils for machinery and food for the crew. Underway training also creates wear and tear on operating equipment, which in turn may increase maintenance demands and costs.

Much of the training for DDG-51 engineering watchstanders currently done underway could be done in port or on simulators at considerable savings. Because of their potentially greater accessibility, simulators may also offer a means to improve engineers' training and preparation for these difficult jobs. It takes many repetitions for watch-

---

[1] Roland J. Yardley, Harry J. Thie, Christopher Paul, Jerry M. Sollinger, and Alisa Rhee, *An Examination of Options to Reduce Underway Training Days Through the Use of Simulation,* Santa Monica, Calif.: RAND Corporation, MG-765-NAVY, 2008.

standers to gain proficiency—repetitions that simulators could provide. By helping watchstanders meet training standards before going to sea, simulators provide a safer way to operate as well as a more efficient way to train.

Given the opportunities a simulator can offer both to reduce costs and increase proficiency, the Director, Assessment Division (OPNAV 81) asked RAND to assess how U.S. Navy surface combatants conduct engineering training and to determine if training efficiencies could be achieved and/or underway time for training could be reduced through a greater use of simulators. Accordingly, this monograph

- describes engineering watch organization and the training and proficiency requirements for engineering watchstanders
- analyzes engineering watchstander performance of training requirements
- assesses currently available engineering simulators and their attributes
- discusses Navy plans for installing embedded engineering simulators on ships
- notes where simulation can best be used for training engineering watchstanders and reduce underway training days
- discusses approaches used for training in the maritime industry and other navies.

To conduct this research, we

- met with subject matter experts from the Afloat Training Groups (ATGs) and engineering experts from a DDG-51 destroyer squadron and discussed how simulators currently contribute to proficiency of engineering officers
- visited and toured the engineering plant and the central control station of a DDG-51 and spoke with shipboard engineers
- went on board ships and questioned both squadron representatives and ships' companies on engineering proficiency training
- met with engineering personnel on board a cruiser with an embedded trainer

- met with engineering experts from the Royal Navy, the Canadian Navy, and the maritime industry
- reviewed reference publications, such as the "Surface Force Training Manual,"[2] "Engineering Department Organization and Regulations Manual,"[3] and other references noted in the bibliography
- identified available engineering simulation technologies and their current use
- met representatives of the Surface Warfare Officer School (SWOS) in Newport, R.I., and examined simulators used to train surface warfare officers (SWOs) for their engineering duties at sea
- met with representatives of the Naval Air Warfare Training Support Division for Surface Systems, which maintains the simulator systems.

In the next chapter, we review DDG-51 engineering watch organization and watchstander training requirements, including engineering training activities that can be done in port. In the third chapter, we explore DDG-51 engineering training requirements and underway days used to accomplish them, and how simulators might help to improve proficiency of engineering tasks. In the fourth chapter, we review currently available simulators and the training that might be performed on them, as well as the Navy's plans for installing embedded trainers onboard DDG-51s. In the fifth chapter, we review the advantages and disadvantages of simulators. In the sixth chapter, we assess the resources and policy changes that would be needed to implement greater simulator training. In the seventh chapter, we summarize our findings and conclusions. Several appendixes supplement the main text.

---

[2] Department of the Navy, COMNAVSURFORINST 3502.1D, "Surface Force Training Manual," Change 1, July 1, 2007.

[3] Department of the Navy, COMNAVSURFORINST 3540.3A, "Engineering Department Organization and Regulations Manual (EDORM)," September 22, 2008b (with change transmittal 1).

CHAPTER TWO

DDG-51 Engineering Watch Organization and Training Requirements

Engineering Watch Organization

The engineering watchstanding organization for surface ships is governed by an engineering department organization and regulations manual (EDORM).[1] The EDORM specifies, by ship class, the minimum number and type of engineering watches that must be stood and the duties of the watchstanders.

The engineering officer of the watch (EOOW) stands watch in the central control station (CCS) and has charge of the engineering watch team. Figure 2.1 shows a notional watchstanding organization of a DDG-51–class ship during underway peacetime steaming. The EOOW is responsible for the safe and proper operation of the ship's engineering plant and for engineering watchstanders' performance.

The CCS is the hub that has the consoles that control the major operations of the engineering plant. The propulsion and auxiliary control (PACC) operator and the electric plant control console (EPCC) operator also work in the CCS. The PACC operator supports the EOOW and controls the main engines and supporting auxiliary equipment. The EPCC operator monitors and controls the ship's electric power plant. The damage control console (DCC), used primarily for starting or stopping the ship's fire pumps and monitoring high-temperature alarms throughout the ship, is also in the CCS. During normal opera-

---

[1] Department of the Navy, 2008b.

Figure 2.1
DDG-51 Watchstanding Organization

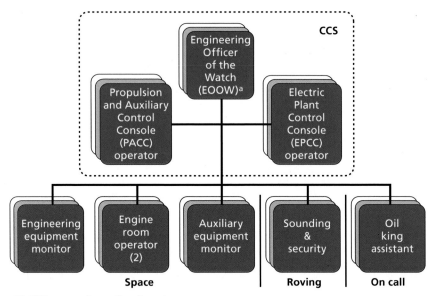

^aEOOWs are senior enlisted engineers.

RAND *MG874-2.1*

tions, the DCC is operated either by the EOOW, the PACC operator, or the EPCC operator.

The EOOW is also supported by watchstanders in the two engine rooms and auxiliary spaces, a roving sounding and security watch, and an on-call oil-king assistant. The forward main engine room (MER 1) and the aft engine room (MER 2) are manned by engine room operators (EROs), who are responsible for the safe and effective operation of equipment in their respective spaces. There is some flexibility in how ships man their engine rooms while underway. For example, officials from the Afloat Training Group, Atlantic (ATGLANT) indicated that an ERO could stand watch in one engine room and an equipment monitor stand watch in the other. The ERO is assisted by an engineering equipment monitor.

There are two main auxiliary spaces, auxiliary space (AUX) 1 and AUX 2. During normal operations, one auxiliary systems monitor

(ASM) mans and monitors the operations of auxiliary equipment in both spaces.

The sounding and security (S&S) watch is a roving patrol that checks the levels of various engineering tanks throughout the ship and makes periodic status reports to the CCS. The oil king assistant takes samples of lube oil, fuel oil, and other equipment fluids for testing of proper operating specifications or contamination, reporting the results to the EOOW.

The total number of personnel on watch varies according to the conditions under which the ship is operating. A ship could be in *cold iron* (plant is not lit off and the ship is drawing electric power from the pier) and operating at minimum manning, *auxiliary steaming* (ship's electric online plant providing power to the ship), with a higher state of manning including an EOOW and EPCC operator, or *underway steaming* status. During underway steaming, there are normally a minimum of eight engineering watchstanders on watch onboard a DDG-51 at any one time. Engineering watches are normally performed by three watch sections, who share watchstanding responsibilities around the clock.

Evaluating Watchstanders

A ship's engineering team is evaluated during an Engineering Operational Certification (EOC). An EOC is a formal evaluation of the ship's engineering team. It is conducted by the ship's immediate superior in command (ISIC) and assisted by ATG. A DDG-51 must pass an EOC at least once every 27 months.[2] During an EOC, engineering watchstanders must effectively perform *engineering evolutions*, that is, actions taken by engineering watchstanders for the normal operation of the engineering plant, and *casualty control drills*, demonstrating their ability to safely operate, control, and restore the engineering plant.

---

[2] The *Surface Force Training Manual* (Department of the Navy, 2007) directs that an EOC must be conducted every 24 months (+3 to −6 months). The range of time for a unit EOC can be 18 to 27 months since the last EOC.

An EOC focuses on and evaluates engineering watchstander proficiency on operations, evolutions, and drills.[3] During an EOC, two of the three ship's watch teams are assessed in their proficiency in performing engineering evolutions and casualty control drills. The third watch team is normally the ship's Engineering Training Team (ETT), which trains and grades the performance of the other two watch sections.

Watch teams must meet proficiency standards. Specifically, each watch team must perform a set of evolutions, with 65 percent being graded as effective. Normally, there are 15 evolutions[4] performed per watch team during an EOC, selected and evaluated by the ETT. Each watch team must also effectively perform engineering casualty control (ECC) drills, with 50 percent of the drills being graded as effective.[5] Eight ECC drills are done per watch team, and at least four of the eight drills must be effective per watch team.

In addition to demonstrating drill proficiency, ships must meet many other challenging requirements during an EOC. Ships must have effective management programs (e.g., managing fuel and lube oil quality, maintaining engineering logs and records) maintenance schedules, meeting minimum equipment requirements to get underway, and being well-maintained and safe to operate. Our research focuses on how well ships meet and maintain proficiency standards and how an engineering simulator can help meet proficiency training demands.

Training Requirements

DDG-51 engineers must be able to safely and effectively operate equipment that is in their charge and be proficient in the watchstations that they stand. Beyond these individual responsibilities, they must work

---

[3]   Department of the Navy, 2007, pp. 2–46.

[4]   *Evolutions* are events performed during the normal operation of the engineering plant and include such actions as aligning equipment for operation, and starting and/or stopping equipment.

[5]   Drills are effective if the watchstanders completed all steps in the procedure as written, in the stated sequence, without deviation; unless deviations were in accordance with approved guidelines.

effectively as a team to operate the engineering plant, control casualties, and restore ship operations after casualties. The EDORM notes that "Watchstanding requires plant operational experience, systems inter-relationship level of knowledge, maintenance and repair expertise, and clear understanding of watch requirements."[6]

Individual Personnel Qualifications

Engineering watchstanders must meet the Personnel Qualification Standards (PQS) for the watch position that they are standing.[7] The PQS delineate the minimum knowledge and skills an individual must demonstrate before standing watch or performing other specific duties necessary for the safe, secure, and proper operation of a ship. An individual must be qualified to operate all equipment under his or her charge before being allowed to stand watch.

Engineering Team Training Requirements

The *Surface Force Training Manual* provides guidance on the type and number of drills and evolutions that engineering watchstanders must perform, as well as the grading criteria and standards that ships must meet to maintain proficiency. DDG-51 engineers are trained to follow exact procedures to bring the plant to an operational status, operate the plant under normal conditions, align and start equipment and take it offline, and perform casualty control when necessary.

These procedures comprise the Engineering Operations Sequencing System (EOSS). The EOSS is a set of written procedures for the normal operation of a ship's engineering propulsion plant. It standardizes operational techniques for watchstanders and casualty control practices. It has two major subsystems: Engineering Operational Procedures (EOP) and Engineering Operational Casualty Control (EOCC).

---

[6]   Department of the Navy, 2008b.

[7]   Department of the Navy, NAVEDTRA 43514-OC, "Personnel Qualification Standard," Naval Education and Training Command, June 2008a.

EOP documents list the steps and systems alignment required for normal engineering plant alignment and operation. EOCC provides watchstanders with step-by-step procedures that must be followed to handle the most commonly occurring casualties. It addresses the actions and communications to recognize the casualties, control the action, and to place the engineering plant in a stable condition. Watchstanders must be proficient in their performance of engineering evolutions and ECC drills.

Evolutions

Evolutions are addressed by the EOSS, but evolutions also come from Planned Maintenance System (PMS) requirements, technical manuals, and locally generated procedures. There are three categories of evolutions: routine, infrequent, and Master Light-Off Checklist (MLOC) (or start-up) procedures. Evolutions require watchstanders to follow specific procedures in aligning equipment for operations, for starting and stopping equipment, and during normal operation of the plant.

Routine evolutions are those that are normally done frequently, i.e., daily or weekly. *Infrequent evolutions* are those that could reasonably be expected to be done during extended operations at sea. *MLOC evolutions* are alignment evolutions performed to maintain proficiency in the safe light-off of an engineering plant. MLOC evolutions and routine evolutions not done more frequently must be done at least quarterly (every three months) for proficiency purposes. Infrequent evolutions must be done annually. Each watch team is evaluated on its performance on 15 evolutions; during an EOC, 65 percent must be performed effectively.[8]

Watchstander proficiency requirements vary by watchstation. Overall, for the entire engineering watch team, there are 101 differ-

---

[8] A routine evolution is graded as effective if the watchstander, without the assistance of ETT or his/her supervisor, conducts all steps in the procedure in accordance with the EOSS User's Guide, as written, in the stated sequence, and without deviation from the applicable EOP, Naval Ship Technical Manual, Planned Maintenance System, manufacturer's or technically correct locally approved procedures. Infrequent or MLOC evolutions use the same criteria, except that watchstanders are allowed a one-time assist from their supervisor in the conduct of the evolution.

ent evolutions, with the number varying by station. We list these evolutions in Appendix A. Figure 2.2 shows the number of engineering evolutions, by watchstation, that must be performed annually per watchstander to maintain proficiency.

Each watch section must complete 339 evolutions annually (some of which are done more than once a year). This means that all three watch sections combined on a DDG-51–class ship must complete more than 1,000 engineering evolutions each year.

Engineering Casualty Control Drills

DDG-51 engineers must be prepared to respond to and control equipment casualties. Equipment may malfunction for various reasons, including personnel errors. Casualty control training ensures that watchstanders make the correct initial response to engineering casualties to reduce further damage to equipment or hazards to the ship and personnel, to prevent additional casualties, and to restore mobility.

Figure 2.2
Annual Number of Evolutions to Maintain Proficiency by Engineering Watchstanders, DDG-51–Class Ships

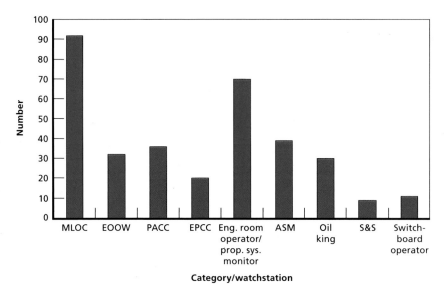

RAND *MG874-2.2*

Watchstanders must commit to memory immediate responses and controlling actions for casualties, in accordance with EOCC, and perform these actions in the correct sequence, from memory. Engineering watch teams are graded on their ability to effectively perform engineering casualty control drills. The reasons a team may be graded as not effective in these drills include

- failure to complete all steps in proper order and when required
- inability to maintain plant control
- committing a safety violation or failure to recognize an unsafe condition.[9]

For the purposes of this study, watchstanders fall into two broad categories, console operators and in-space watchstanders. The EOOW, EPCC, and PACC operators perform their required actions at standard consoles. Their training can be easily augmented by use of simulators. Because the in-space watchstanders act on a variety of controllers, valves, motors, etc. in various spaces throughout the ship, their training is not easily augmented by simulation.

There are three categories of engineering casualty control drills. Category I drills are casualties with high risk and/or those that occur most frequently. Category II drills are casualties with a moderate risk or those that occur frequently. Category III drills involve low risk or infrequent occurrence. We list these drills in Appendix A. Category I drills must be effectively done by each watch team every three months, Category II drills every six months, and Category III drills, annually.

There are a total of 40 ECC drills, which as noted above, are repetitive. During an EOC, watchstanders are assessed on their performance of Category I and II drills.

---

[9]   Department of the Navy, COMNAVSURFOR Instruction 3502.1C, "Surface Force Training Manual," January 1, 2006a.

Performing Drills in Port

Our previous research has shown that many engineering casualty control drills can be done in port, with no simulation required.[10] To conduct casualty control drills in port, the plant must be hot (main engines lit off) and all watchstations manned. Training in port offers a less expensive and viable option to increase the proficiency of engineering watchstanders before going to sea. Fuel is expended during in-port training, but to a much lesser extent than in underway operations. Table 2.1 lists all engineering casualty control drills and denotes those that can be done in port.

Table 2.1
Engineering Casualty Control Drills

Drill	Name	Can Be Done in Port
Category I		
MMFOL	Major fuel oil leak	X
MBGTM	Class "B" fire in GTM module	X
MLLOPR	LOSS L/O pressure main reduction gear (MRG)	
MHBRG	Hot bearing in MRG	
MOSGG	Overspeed SSGTG	X
MBGGM	CLASS "B" fire in gas turbine generator (GTG)	X
MCBF	Class "B" fire in main space	X
MCFED	Class "C" fire electrical distribution system	X
MMF	Flooding in main space	X
Category II		
MLFOP	Loss of fuel oil pressure	X
MGGS	Gas generator stall	X
MLPACC	Loss of PACC	X

---

[10] Yardley et al., 2008.

Table 2.1—Continued

Drill	Name	Can Be Done in Port
MLSCU	Loss of shaft control unit	X
MLCRP	Loss of pitch control	
MLHOL	Hydraulic oil leak	
MLLOL	Major leak MRG loss of lube oil	
MNVRG	Noise/vibration/MRG/shaft	X
MHLSB	Hot line shaft bearing	X
MLHOP	Loss of controllable pitch propeller hydraulic oil pressure	
MHBGTG	Hot bearing in GTG	X
MGHIT	High turbine inlet temperature GTG	X
MLGGO	Low lube oil pressure to GTG	X
MPSFG	Post shutdown fire in GTG	X
MPSFR	Post shutdown fire in Redundant Independent Mechanical Start System (RIMSS)	X
MCCFG	Class "C" fire in GTG	X
MCCFS	Class "C" fire switchboard	X
MLSC	Loss of steering control	X
MCASF	Gas turbine cooling air system failure	X
Category III		
MLPTO	Loss of lube oil pressure GTM	X
MEPTV	Excessive power turbine vibration GTM	X
MGGOS	Gas generator overspeeds GTM	X
MHTIT	High gas turbine inlet temperature GTM	X
MLPLA	Loss of power lever actuator GTM	
MPTOS	Power turbine overspeeds GTM	X
MPSFP	Post-shutdown fire turbine case GTM	X

Table 2.1—Continued

Drill	Name	Can Be Done in Port
MHST	High propulsion shaft torque	X
MLEPC	Loss of electric plant control	X
MFZDB	Electrical fault zonal main distribution bus	X
MNVGG	Unusual vibration noise in GTG	X
MLCWS	Loss of chilled water system	X

The consoles in CCS provide watchstanders with an audio and/or visual alarm when a casualty or out-of-limit condition occurs. CCS watchstanders' actions in response to casualties may include communicating with in-space watchstanders, stopping online equipment, isolating the equipment, and starting offline equipment to maintain the required capability.[11]

Engineering Watchstanders

We examined data provided by ATGLANT to gain a better understanding of the grades of the personnel, officer or enlisted, who stand engineering watches. These data, and discussions with senior surface warfare officers (SWOs) and interviews with engineering training subject-matter experts from the Afloat Training Group, Pacific (ATGPAC) and ATGLANT, indicate that few officers stand engineering watches (Figure 2.3). Senior enlisted personnel are those in the grades of E-7 through E-9. Enlisted are those in grades E-1 through E-6.

Although enlisted personnel are primarily responsible for engineering watches, our earlier research indicated simulators are used pri-

---

[11] A DDG-51 engineering plant has considerable redundancy. That is, it has offline equipment that can be started to maintain needed capabilities. For example, there are three gas turbine generators, only two of which are normally online. If a casualty were to occur to one of the generators, the EPCC operator would isolate the affected generator, start the offline generator, and parallel it with the generator that remained online.

Figure 2.3
DDG-51–Class Engineering Watchstanders, by Position, by Grade, for
Atlantic-Based Ships Undergoing ULTRA-E and EOCs, 2005–2007

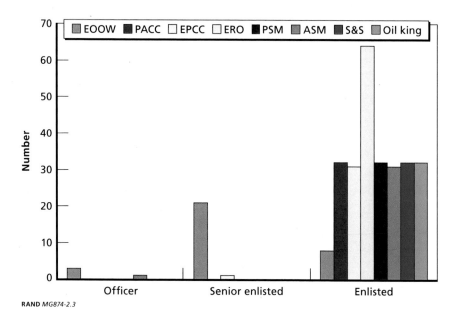

RAND *MG874-2.3*

marily to train officers for watchstanding. Indeed, few officers stand engineering watches.

Overall, the data indicate that more than 95 percent of all engineering watchstanders were enlisted personnel. Yet existing simulator resources, including desktop and console engineering simulators, are used extensively to only train officers—prospective department heads and EOOWs at Newport, R.I. These valuable training resources are not readily available for use by enlisted personnel who stand the watch. We will revisit this question when we consider simulators as a training resource. In the next chapter, we consider how ship engineers perform in unit-level training for engineering evolutions and casualty control.

CHAPTER THREE

Engineering Training Performed by DDG-51s During Unit-Level Training

After ships return from deployment, they normally undergo a maintenance period and then begin unit-level training (ULT). There is usually a turnover of personnel after deployment, with some seasoned engineers rotating off the ship and new personnel reporting onboard. A ULT period allows a ship's crew to become proficient in operating the ship and its systems.

During the initial phase of ULT, ship's engineering teams are first assessed on their engineering proficiency, and then a tailored training program is developed to address training deficiencies. The assessment period is called a Unit Level Training Readiness Assessment–Engineering (ULTRA-E). The ULTRA–E is conducted by the ISIC and assisted by ATG. After the ULTRA-E, a combination of in-port and underway training (or mobility–engineering [MOB-E] periods) is scheduled to train both the ship's training teams and the crew. At the end of tailored MOB-E training, and normally after having demonstrated the proficiency necessary to succeed, ships undergo an EOC.

Figure 3.1, drawn from our previous work, shows the number of exercises done underway in each mission area for DDG-51–class ships underway in ULT, for each mission area in calendar year (CY) 2004. Engineering training requires much practice, repetition, and focus and consumes time and resources. The red bar represents the number of engineering casualty control exercises done underway, which far exceeds the number of exercises done in any other mission area The number of engineering evolutions done is not reflected in Figure 3.1.

Figure 3.1
Number of Exercises Done Underway in Unit-Level Training, DDG-51–Class Ships, by Mission Area, CY 2004

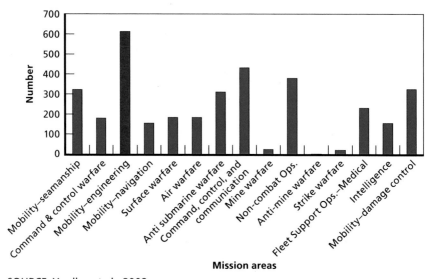

SOURCE: Yardley et al., 2008.
RAND *MG874-3.1*

A great deal of effort is dedicated to improving the engineering casualty control proficiency of engineering watchstanders. It normally takes more than 100 engineering casualty control drills before watch teams gain proficiency to meet the minimum standard for drills of 50-percent effectiveness.

Casualty control drills must also be repeated for watchstanders to *maintain* proficiency. To maintain proficiency, 85 drills must be "effectively" performed by each watch team annually. Ships must plan and perform extra drills to make up drills that are graded as not effective. As a result, ships must perform at least two repetitions to meet the standard. Annually, at least 170 repetitions per team must be scheduled or 510 total for a ship (170 repetitions for each of the three ship's watch teams) to meet the ship's readiness requirements.

In this chapter, we examine unit-level training in four ways. First, we assess data on watchstander performance of evolutions and drills, including how many repetitions and underway days it takes ship crews to meet proficiency standards. Second, we assess the most common causes of failure in training. Third, we examine the steps that must be performed by the Engineering Training Team that evaluates watch-standers. Fourth, we broadly address how simulators may help improve training.

Performance Data

To examine the effects of training demands on underway days, we sought proficiency data for ships undergoing ULT from ATGLANT and ATGPAC. The data had some missing elements. To ensure the accuracy of our analysis, we used only complete data for this research. This enabled us to track 35 Atlantic ships and 10 Pacific ships from ULTRA-E through EOC.

Performance on Evolutions
In both ULTRA-E and EOC, ships must effectively perform 65 percent of evolutions. If ships do not meet the 65 percent standard during the ULTRA-E, the qualifying standard is increased to 75 percent. If ship engineers achieve 75 percent proficiency during the first three weeks of MOB-E Tailored Ship Training Availability (TSTA) training periods, then they could potentially validate an EOC (if other EOC standards are also met). Though ships on the East and West coasts perform the MOB-E TSTA training progression slightly differently, the standards for passing are the same on both coasts.

ATGLANT collects training data for ships from East Coast homeports. Data for 35 ships were collected and used for this study. Figure 3.2 shows the progression of effectiveness for evolutions for East Coast based ships.

Ships that demonstrate a combined effectiveness rate of 65 percent for evolutions and 50 percent for drills at ULTRA-E are considered complete from a proficiency perspective and are ready for tasking. They must also demonstrate effective programs and sound material condition to achieve MOB E certification. Among the 35 Atlantic ships for which we have ULTRA-E data, the initial effectiveness rate averaged

Figure 3.2
Percentage of Engineering Evolutions Graded as Effective by Atlantic-Based DDG-51s, CYs 2005–2007

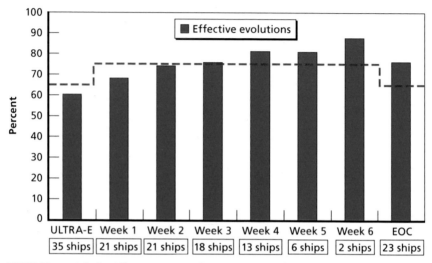

NOTE: The red dashed line represents the minimum proficiency requirement for passing the engineering training event.
RAND MG874-3.2

just over 60 percent for engineering evolutions. Should a ship not pass in ULTRA-E, it begins week 1 of MOB-E TSTA. Of the 35 Atlantic ships in ULTRA-E, we had complete data on 21 that did not meet proficiency standards and proceeded to MOB-E training.[1]

Although some East Coast ships spent up to six weeks in MOB-E/TSTAs, the data provided by ATG San Diego indicated that West Coast ships spent approximately three weeks undergoing engineering training.[2] Like East Coast ships, at ULTRA-E the West Coast ships evolution proficiency started with an effectiveness rate around 60 per-

---

[1] The data provided by the Afloat Training Groups included underway dates and the number of evolutions and drills attempted and graded effective. While ULTRA-E data made up the most complete data set, dates were missing for several ships in MOB-E training. We include only those ships for which we had complete data.

[2] ATGPAC provided data for only calendar year 2007, for ten ships undergoing ULT. We do not know if West Coast ships used more or less time in past years.

cent. The West Coast ships also gained proficiency at about the same rate as the East Coast ships.

The data from both coasts shows that crew performance of evolutions starts below the standard, but the crews become proficient quickly and tend to stay proficient.

Engineering Casualty Control Drill Proficiency

By contrast, both East and West Coast ships are less proficient at ECC drills. As discussed earlier, drills are generally more complex, require a whole team response, and must be completed from memory. The ECC drills also have a lower passing standard to achieve certification: 50 percent effectiveness in ULTRA-E and EOC and 60 percent during the first three weeks of MOB-E/TSTA. As Figure 3.3 shows, the 35 Atlantic ships achieved only 35 percent proficiency in their ULTRA-E drills. Again, we had complete data on 21 of the 35 ships that had to move on to MOB-E training.

Figure 3.3
Percentage of Engineering Drills Graded as Effective by Atlantic-Based DDG-51s, CYs 2005–2007

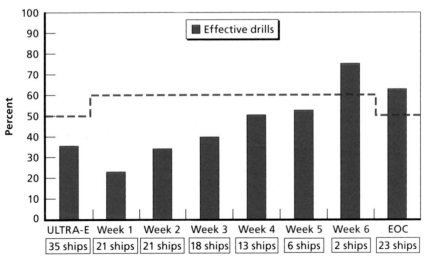

NOTE: The red dashed line represents the minimum proficiency requirement for passing the engineering training event.
RAND MG874-3.3

Figure 3.4 shows that West Coast ships achieved only 25 percent effectiveness in ULTRA-E, and that eight of the ten ships entered MOB-E training. All West Coast ships achieved 60 percent effectiveness on these drills within three weeks of MOB-E/TSTA training, but again, we have a smaller data set for West Coast ships.

We next looked at the data to examine how many drills it took and how many underway days were used before ships first achieved the 50-percent standard.[3] On average, West Coast ships needed 100 drills to meet the 50-percent standard, and East Coast ships needed 138. The average number of underway days used were 13 and 20 respectively, for West and East Coast ships.

Some ships qualified (met the proficiency standards) for the EOC at the ULTRA-E, or the initial assessment. These ships used a reduced

Figure 3.4
Percentage of Engineering Drills Graded as Effective by PAC-Based DDG-51s, CY 2007

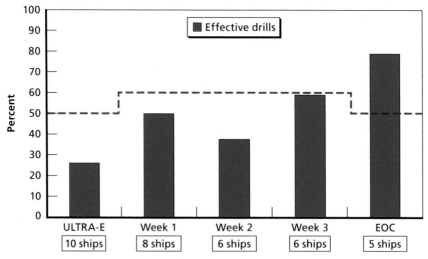

NOTE: The red dashed line represents the minimum proficiency requirement for passing the engineering training event.
RAND MG874-3.4

---

[3] A normal casualty control drill set for an engineering watch section consists of eight drills.

number of drills and ATG-assisted training underway days to achieve EOC. We combined the drill and underway data to produce Figure 3.5. The ships that passed at ULTRA-E are indicated in the blue-shaded box in the figure. We examined the employment schedules of the ships that qualified at ULTRA-E to determine, from a scheduling view-point, what circumstances supported the early qualification. We found that all of the ships that qualified early had at least ten underway days within the preceding two months leading up to the ULTRA-E. We found that other ships that did not qualify at ULTRA-E also had at least ten underway days prior to the ULTRA-E. Destroyer-class squad-ron (DDGRON) authorities stated that other factors—e.g., command attention to engineering readiness, personnel experience and turnover, training, material condition of ship—could account for differences between ships.

It takes many repetitions before watch teams become proficient in effectively conducting casualty control drills. In addition, many

Figure 3.5
Underway Days and ECC Drills Needed to First Meet 50 Percent
Effectiveness Standard for DDG-51s in Unit-Level Training, East
and West Coast Ships

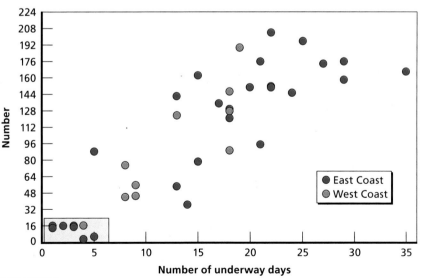

underway days (resources) are used to get ship's engineering teams to the point of first beginning to meet the fleet standard for engineering casualty control drill proficiency. Crews use underway time to train in all mission areas, not just engineering.

We also note that the number of underway days used in ULT for East and West Coast ships is different. East Coast ships in ULT are tasked to support the fleet commander to a greater extent than are West Coast ships. They are used more—i.e., they have greater demands placed on them—when underway for ULT than are their West Coast counterparts. We spoke with West Coast scheduling authorities who indicated that they try to "protect" ships in ULT from other demands that would affect their training. Therefore, there are differences in the demands placed on East and West Coast ships that may impact the time available to conduct engineering training.

Table 3.1 is a summary table of the data contained in Figure 3.5. The data represent the ECC drill proficiency of two watch teams per ship. For most West Coast ships, it took a total of 100 drills before the watch teams first achieved the 50-percent effectiveness rate, which equates to slightly more than six drill sets per watch team. East Coast ship watch teams required more than eight drill sets per watch team before they first met the proficiency standard.

Altogether, 73 percent of drills in our data were Category I drills. Category II drills made up 24 percent of the total drills; Category III, 3 percent. Most of these drills can be done pierside while onboard, or on an engineering simulator. Doing drills pierside or on a simula-

Table 3.1
Repetitions and Underway Days Needed to Meet the 50-Percent Standard

DDG-51s in ULT, 2005–2007	East Coast	West Coast	Both
Number of ships	30	10	40
Average number of drills to meet standard (not including ships in the blue box)	138	100	126
Average number of underway days to meet standard (not including ships in the blue box)	20	13	18

tor can serve to increase the proficiency of engineering watchstanders before ships go to sea.

Causes for Failure

We next examined the reasons for casualty control drill failure and which watchstanders were responsible. These have implications for simulator training: Simulators could perhaps help boost proficiency for CCS watchstanders, but there are no current DDG-51 simulators capable of training those who stand watch in the engine rooms and auxiliary spaces.

We previously noted the broad reasons why engineering drills are graded as not being effective. The ATG has developed critique sheets that further describe specific watchstander actions (or inactions) that cause a drill to be graded as ineffective. Possible reasons include the following:

A. Steps were conducted out of sequence.
B. Steps were missed.
C. Did not use procedure.
D. Steps were performed improperly.
E. Insufficient knowledge to conduct evolution.
F. Did not obtain permission from a supervisor for a step.
G. Caused a loss of plant control.
H. Failed to report/take action on alarm.
I. Failed to recognize material discrepancy.
J. Failed to recognize documentation problem.
K. Failed to report material discrepancy.
L. Failed to report documentation discrepancy.
M. Self-simulated actions.
N. Inordinate delay in actions.
O. Did not wear personal protective equipment.
P. Did not recognize unsafe action.
Q. Committed general safety violation.

We combined the reasons for failure into four broad categories and identified where failures occurred in the data we examined. Figure 3.6 summarizes the results for Atlantic-based ships, showing the number of failures attributed to each category and area.

Of the drills that were graded as not effective, we found that CCS watchstanders failed 71 percent of drills and that in-space watchstanders failed 64 percent of drills. Many drills were failed by both. In short, both CCS and in-space watchstanders have challenges in effectively performing ECC drills.

Although the use of an engineering simulator can help increase the proficiency of CCS watchstanders in learning their ECC drill actions, the actions of in-space watchstanders must also be improved. The engineering team needs to perform well as a unit to succeed in drills. In casualty control drills, in-space watchstanders are required to

Figure 3.6
ATGLANT Assessment of DDG-51 ULTRA-E and EOC ECC Drill Failures, Atlantic-Based Ships, CY 2007

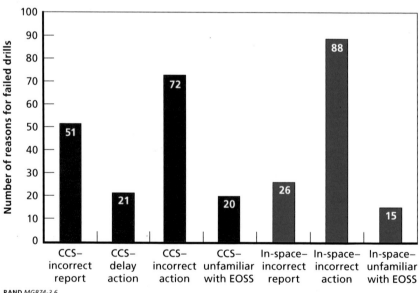

RAND MG874-3.6

know the exact location and operating procedures for valves, controllers, and equipment in response to casualties. Much of this training can be done in port and requires repetition and practice. Although an engineering simulator cannot assist with an in-space watchstander's performance, increased in-port training time can help produce improved performance.

Engineering Training Team

In addition to having proficient watch teams, each ship must have its own training capabilities. Each ship's Engineering Training Team must be found proficient in evaluating an engineering watch team's performance of evolutions and casualty control drills. To run a casualty control drill set on engineering watchstanders, the ETT must

- determine who will be on watch and the drills or evolutions that will be run during a watch
- schedule evolutions and drills so they will not conflict with other shipwide events or affect tactical mobility of the ship
- request and get approval from the commanding officer to run the drills
- meet as a team, brief the drill set, decide how the drills will be imposed and by whom, and decide how members of the ETT will communicate with each other throughout the drills
- complete a safety walkthrough of engineering spaces before starting the drills and evolutions
- ensure that heat stress conditions in the plant support the conduct of training
- run the drills and evolutions; communicate, coordinate, and integrate actions during the drill
- critique and grade the watchstanders
- meet together, compare notes, and critique drill performance as a team
- debrief the watchstanders on their performance
- record the evolutions and drills completed.

As noted earlier, the third watch team on a ship typically trains the other two, that is, it comprises the ETT. Because it alone can train and critique the other two watch teams, there is no team that evaluates their performance. Yet whether this is the best engineering watch team is not clear. ATGLANT and ATGPAC senior trainers told us that because it is in a ship's interests to pass the EOC, the ship may present the best watchstanders in sections one and two for evaluation.

How Can These Challenges Be Addressed by Simulators?

Our data indicate that many drills are not evaluated during ATG-assisted training events (for example, as noted above, 73 percent of drills for ATG-assisted training were CAT I drills). In addition, many watchstanders never get the opportunity to perform and be evaluated, because only two of the three watch teams onboard a ship are tested. The data also indicate that a large number of ship's engineering teams do not initially meet proficiency standards and must continue their training past ULTRA-E and into MOB-E TSTA periods. That is, many crews get underway without meeting the proficiency standards for handling engineering casualties aboard their ship.

The commercial maritime industry and navies around the world use ship engineering simulators to overcome these problems, both to increase the proficiency and readiness of a crew before it goes to sea and for continuation training. (See Appendix C for more information on how these organizations use simulators for engineering training.)

A shore-based simulator would allow for more of a ship's company to be trained on each drill as well as for more drills to be practiced with fewer time constraints. Engineering watchstanders on a ship at sea are primarily concerned with maintaining mobility, which entails performing required maintenance, troubleshooting, and repairs. Operating and maintaining a DDG-51 requires a significant level of effort by the entire engineering department.

We view the passing rate—that 50 percent of drills performed must be effective—as a low standard. Simulators on shore can provide the opportunity to increase watchstanders proficiency above the standard before going to sea and provide a ready and accessible training resource that is currently not available for DDG-51–class ships at fleet concentration areas.

In the next chapter, we describe the engineering simulators that are available and could be used by DDG-51 engineers, and the Navy's plan for installing embedded trainers on DDG-51s.

CHAPTER FOUR

What Simulators Are Available?

Engineering training requirements, as we saw in Chapter Two, are quite extensive. Some ships, as we saw in Chapter Three, require many repetitions and underway days before ship engineers first meet the 50-percent effectiveness standard in their drills. Of course, if only half the drills are effective, then the other half are not. Put another way, the standard is low and many ships need time to meet it. An engineering simulator could help engineering watchstanders meet and exceed proficiency standards.

SWOS Engineering Simulators Can Be Used to Train CCS Watchstanders

The U.S. Navy Surface Warfare Officers School uses two simulators to train officers for shipboard engineering duties. It uses a DDG-51 desktop simulator (Device 19G4A) and a console trainer (Device 19G4) to provide DDG-51 Machinery Control System (MCS) training for prospective engineering department heads and EOOW students. The training at SWOS is an eight-week class of instruction. Engineering students receive engineering theory and fundamental training in their first three weeks of class and then proceed to class specific (DDG-51 engineering plant) training. The students are introduced to the desktop console trainer and learn to align, start, and stop engineering systems on the trainer. After gaining proficiency on the desktop trainer, students then move to the console trainers for individual and team training. At the end of the training, ATG assessors evaluate the students' performance on the consoles. Below we describe the operation and attributes of the desktop and console simulators.

DDG-51 Desktop Simulator

Instructors use the trainer to meet course objectives related to gas turbine engine principles, gas turbine engine watchkeeping, and operation of the DDG-51 Class MCS consoles. Students use the trainers to execute normal (EOP) and casualty control (EOCC) operations. The trainer mimics shipboard equipment and responds to console operator inputs in the same manner as the actual DDG-51 shipboard MCS consoles.

The desktop simulator provides 3-D graphical user interface representations of the MCS consoles. The controls and switches can be accessed through the keyboard and mouse clicks. These 3-D representations are known as "virtual" MCS consoles. The desktop operates on a two-monitor display.[1] During normal operation, it displays the MCS panel on the left monitor and the EOSS program on the right. The entire EOSS program is available in the software (in Adobe Acrobat format). For normal plant operation, to perform an evolution during a training scenario, clicking on a console's manual will bring up that manual for use. Step-by-step procedures are provided in the EOP to conduct the evolution, and the steps must be followed exactly. All of the consoles have both an EOCC and an EOP manual. The ship control console and the damage control console have one EOP only. Figure 4.1 depicts the dual-screen display of the desktop trainer.

Desktop Simulator Scope of Operation. The desktop simulators (student stations), when connected together via a local area network, can operate in the same manner as the full-size console, except with "virtual" MCS rather than hardware consoles. At the SWOS, one of the desktop simulators is used as an instructor operating station, and the others are set up as MCS consoles.

---

[1] The desktop console can also operate on a single monitor. The monitor can be used to display the MCS console and the trainee can use a hard copy of the EOP or EOCC manual (if a second monitor is not available).

Figure 4.1
DDG-51 Gas Turbine Propulsion Plant PC-Based Trainer (19G4A) at SWOS

SOURCE: Naval Air Warfare Center Training Systems Division, "DDG-51 Gas Turbine
Propulsion Plant Training Devices (19G4/19G4A)," briefing, undated.
RAND *MG874-4.1*

Trainee Interface. The trainee interface consists of the four CCS
MCS consoles (PACC, EPCC, DCC and EOOW), MER 1 and 2,
shaft control units (SCU 1, SCU 2), and the bridge control unit (BCU)
portion of the ship control console (SCC) that is located on the ship's
bridge. We describe below each of these seven virtual consoles and the
operations their software simulates.

Damage Control Console (DCC). Control station for the six fire
pumps, remote-operated firemain valves, washdown countermeasures valves and Vertical Launching System space secondary drainage
valves.

Propulsion and Auxiliary Control Console (PACC). Control station for the four LM2500 gas turbine engines (Gas Turbine Modules
1A, 1B, 2A, 2B), port and starboard shafts, associated propulsion and
auxiliary equipment. The digital Ship Speed Indicator is mounted on
top of the PACC.

Electric Plant Control Console (EPCC). Control station for the three Allison 501-K34 gas turbine engines, AC ships' service gas turbine generators (SSGTGs) and bus tie circuit breakers in the electric power distribution system.

Engineering Officer of the Watch/Logging Unit (EOOW). Information display station for the engineering officer of the watch. The EOOW logging unit also contains the bell logger printer, a dummy fuse assembly, and a dummy bubble memory unit.

MER 1 Shaft Control Unit (SCU 1). Control station for the two LM2500 gas turbine engines (GTMs 1A and 1B) in MER 1, the starboard shaft, associated propulsion, and auxiliary equipment. SCU 1 includes the devices that allow manual control of engine speed and propeller pitch for the starboard shaft.

MER 2 Shaft Control Unit (SCU 2). Control station for the two LM2500 gas turbine engines (GTMs 2A and 2B) in MER 2, the port shaft, associated propulsion, and auxiliary equipment. SCU 2 includes the devices that allow manual control of engine speed and propeller pitch for the port shaft.

SCC Bridge Control Unit (BCU). During normal operation, the SCC BCU controls all LM2500 gas turbine propulsion engines (GTMs 1A, 1B, 2A and 2B) by using a port and standard throttle levers for shaft speed. The control of the turbine propulsion engines can be transferred to the PACC or the SCUs if a failure occurs at the bridge. For training purposes, the washdown countermeasures (WDCM) panel can be accessed from the BCU remotely. The WDCM panel is located on the bulkhead behind the SCC on the real ship. Moreover, a mimic panel of the bridge alarms can be accessed and used to provide general, chemical and collision alarm sounds.

DDG-51 Console Trainer

Device 19G4 Gas Turbine Propulsion Plant Trainer Description. The Device 19G4 Gas Turbine Propulsion Plant Trainer (GTPPT) is a MCS simulator for the DDG-51–class ship. The trainer provides students with a realistic simulation that mimics actual shipboard equipment and responds to console operator inputs in the same manner as the actual DDG-51 shipboard MCS consoles. The GTPPT is used to provide interactive, real-time training for prospective engineering department heads and EOOWs.

The console trainer consists of a central processing system with associated software and has an instructor control station. The students operate the console just as they would operate their consoles in CCS. The consoles control the operation of the trainer during normal and casualty control operations. The central computer runs a math model that allows the trainer to simulate normal and casualty operations in DDG-51 propulsion, electrical, and auxiliary systems.

The console trainer and the desktop console trainer use the same software. This software has been upgraded and refreshed to accurately represent the MCS for all three variants of the DDG-51 class (Flights I, II, and IIA), i.e., the software has been upgraded as the DDG-51–class MCS has evolved. Figure 4.2 displays the GTPPT that is in use at SWOS.

Device 19G4 Scope of Operation. The Device 19G4 is controlled by a central processing system (CPS) computer, located at the instructor operator station, that runs a mathematical model of the DDG-51 engineering plant. The mathematical model is based on the actual physics of the gas turbine propulsion plant, and it incorporates the interrelationships between major systems and equipment, as well as the interactions with supporting auxiliary and electrical systems. The CPS computer generates the outputs that are displayed to the students and are affected in real time by student (or instructor) inputs. This allows numerous possibilities for casualty simulations, as well as for normal or abnormal plant operations. The instructor may inject multiple casualties, failures, or abnormal conditions into the training scenario at any time.[2]

Both the desktop and console simulators can be put in freeze mode. This mode freezes the application at its current state within a scenario and pauses the training session. While the program is in freeze, the trainee can be given instruction on his actions (or inactions) and the implications of these on plant operations. The program can

---

[2] Naval Air Warfare Training Systems Division, Training System Support Document (TSSD), Orlando, Fla., July 1, 2008.

Figure 4.2
DDG-51 Gas Turbine Propulsion Plant Trainer (19G4) at SWOS

SOURCE: Photo courtesy of Surface Warfare Officers School Engineering Department, Newport, R.I.
RAND *MG874-4.2*

then resume from where it was frozen. As an alternative, the program or scenario can be cancelled and restarted.

SWOS Simulators Can Be Used to Train DDG-51 CCS Watchstanders

Although they are not currently available to the fleet, the SWOS engineering simulators could be a valuable training resource for DDG-51–class ships. Students can learn to bring the plant to full operation from a cold-iron status, i.e., from MLOC, on the desktop trainer, before progressing to the console trainers for performing engineering evolutions, operational procedures, and casualty control drills. Because the desktop and console trainers respond to actions as would an actual DDG-51 plant, students receive PQS credit for operating the plant, just as if they had performed the actions onboard the ship. To qualify as an EOOW, a student must perform ERO, EPCC/Switch-board Operator (EPCC/SWBD OP), PACC and EOOW watchstanding tasks. As Figure 4.3 indicates, students can achieve 46 percent of DDG-51 EOOW 300-level PQS tasks on the SWOS simulator. We list the 300-level PQS line items that can be satisfied by the simulator in Appendix D.

Training and testing of effectiveness for most casualty control drills can also be done on a simulator. Altogether, as Table 4.1 shows, simulators can provide training and testing for 35 of 40 casualty control drills, including all Category III drills that the CCS watchstanders are responsible for learning. We also indicate those drills that can be done onboard the ship in port without a simulator.

Figure 4.3
DDG-51 EOOW (NAVEDTRA 43514-0C) 300-Level Total Tasks and Tasks Fulfilled by SWOS Simulator

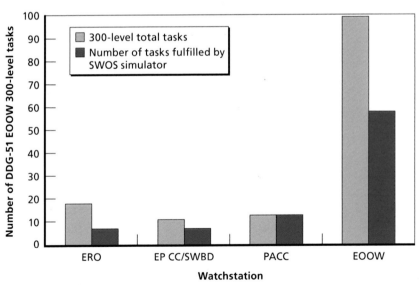

Table 4.1
Engineering Casualty Control Drills That Can Be Evaluated in a DDG-51 Console Trainer

Drill	Name	Can Be Done in Port	Can Be Done via Console Trainer
Category I			
MMFOL	Major fuel oil leak	X	X
MBGTM	Class "B" fire in GTM module	X	X
MLLOPR	Loss of L/O pressure in MRG		X
MHBRG	Hot bearing in MRG		X
MOSGG	Overspeed SSGTG	X	X
MBGGM	Class "B" fire in GTG	X	X
MCBF	Class "B" fire in main space	X	
MCFED	Class "C" fire in electrical distribution system	X	X
MMF	Flooding in main space	X	X
Category II			
MLFOP	Loss of fuel oil pressure	X	X
MGGS	Gas generator stall	X	X
MLPACC	Loss of PACC	X	
MLSCU	Loss of shaft control unit	X	
MLCRP	Loss of pitch control		X
MLHOL	Hydraulic oil leak		X
MLLOL	Major leak MRG loss of lube oil		X
MNVRG	Noise/vibration/MRG/shaft	X	X
MHLSB	Hot line shaft bearing	X	X
MLHOP	Loss of controllable pitch propeller hydraulic oil pressure		X
MHBGTG	Hot bearing in GTG	X	X

Table 4.1—Continued

Drill	Name	Can Be Done in Port	Can Be Done via Console Trainer
MGHIT	High turbine inlet temperature	X	X
MLGGO	Low lube oil pressure to GTG	X	X
MPSFG	Post-shutdown fire in GTG	X	X
MPSFR	Post-shutdown fire in RIMSS	X	X
MCCFG	Class "C" fire in GTG	X	X
MCCFS	Class "C" fire switchboard	X	
MLSC	Loss of steering control	X	
MCASF	Gas turbine cooling air system failure	X	X
Category III			
MLPTO	Loss of lube oil pressure	X	X
MEPTV	Excessive prop turbine vibration	X	X
MGGOS	Gas generator overspeeds	X	X
MHTIT	High gas turbine inlet temperature	X	X
MLPLA	Loss of power lever actuator		X
MPTOS	Power turbine overspeeds	X	X
MPSFP	Post-shutdown fire turbine case	X	X
MHST	High propulsion shaft torque	X	X
MLEPC	Loss of electric plant control	X	X
MFZDB	Electrical fault in zonal main distribution bus	X	X
MNVGG	Unusual vibration noise in GTG	X	X
MLCWS	Loss of chilled water system	X	X

Simulators Can Help Engineers Increase Proficiency

In sum, desktop and console trainers can complement shipboard training for CCS watchstanders. They can provide focused sessions for watchstanders to recognize casualties, practice actions, make reports, and increase familiarity and proficiency with ECC procedures. Simulators are a more efficient way to train because they require less manpower and time to run drills, allow for increased repetition and practice, and offer the opportunity to "freeze" drills and provide watchstanders immediate instruction on incorrect actions.

During casualty control drills (and actual casualties) watchstanders must respond to the symptoms of the casualty; take controlling, immediate and supplemental actions from memory; and restore the plant to a normal operating configuration. This training is time consuming and repetitive because all members of the watch section must function effectively as individuals and as a team and because learning and memorizing the correct actions takes repeated exposure. Watchstanders need repeated exposure and repetition to increase their effectiveness in performing these drills. An engineering console trainer has great value in improving CCS watchstanders' casualty control drill proficiency.

Simulators can provide better training than underway training because they make it possible to repeat a drill several times in a brief training period. In addition, a simulator allows the trainee to repeat just a portion of a drill (e.g., the very start of a drill where many actions take place simultaneously) even more times in a brief drill period. We believe that repetition is the foundation of ECC drill learning— the more repetition the better. Navy instructor manuals imply that actions most often repeated are best remembered, and this idea is the basis of performing drills and practices.[3] Moreover, it has been proven that students learn best and retain information longer when they have meaningful practice and repetition. Frequent and rapid repetition is an improved method of training.

---

[3] Department of the Navy, "Navy Instructional Theory," NAVEDTRA 14300, August 1992.

Figure 4.4 illustrates the use of a console trainer for trainees at SWOS. If console trainers were located at FCAs, ship's CCS watchstanders (EOOW, EPCC, and PACC operators) could use the consoles to increase and meet proficiency standards with the goal of achieving these standards before ships go to sea.

Simulators can allow all CCS watch sections to increase their proficiency, including the third onboard watch team that does not currently receive proficiency training because of its responsibilities in training the other two watch sections. Simulators can also offer training opportunities when a ship is undergoing maintenance. Finally, simulator use, by reducing underway days needed for engineering training, can save fuel dollars and wear and tear on ship's equipment.

As noted, simulators are located only at SWOS, where they are used to train officers. Engineering watch teams are made up of enlisted

Figure 4.4
DDG-51 Engineering Student Being Observed by a SWOS Instructor on a Console Trainer

SOURCE: Photo courtesy of Surface Warfare Officers School Command.
RAND *MG874-4.4*

personnel. Locating engineering simulators where enlisted personnel could benefit from the training capability would improve watchstanders' proficiency. Further, a simulator could help watchstanders to meet proficiency standards before ships go to sea, and allow underway training to be a time to fine tune their watchstanding skills.

Simulators embedded on ships offer opportunities to increase training opportunities. The Navy recognizes the value of an embedded engineering training capability. We next discuss the Navy's plans to use embedded training.

Plans for Backfitting and Use of Engineering Embedded Trainers

The newest DDG-51s (i.e., DDG-96 and newer) have an embedded engineering trainer. This trainer allows the engineering consoles in CCS to be put into a training mode. This device can be used to increase the proficiency of watchstanders in performing casualty control procedures. It allows training to be done onboard, on the ship's own equipment—either in port or underway. The Navy also has a plan to backfit older DDG-51–class ships—i.e., DDG-51 through DDG-95—with an embedded trainer. The backfit plan will begin when DDG-51s commence their midlife upgrades in fiscal year (FY) 2010.

Figure 4.5 displays the current and estimated modernization plan and indicates the number of DDG-51s in the FCAs of Norfolk and San Diego that will be without an embedded trainer. Many DDG-51s do not have an embedded training capability now, and it will take until approximately 2025 to provide all destroyers with such a trainer. In Chapter Six, we consider the resources needed to purchase and sustain these trainers, and the savings that might result from their use.

The Navy is taking steps to do more engineering assessments and certifications in port for ships that have an embedded training capability. Thirteen cruisers have Smartship modifications, including an onboard trainer (OBT) for CCS watchstanders. The onboard trainer provides an extra console for the EOOW, PACC and EPCC operators, and is used for practice and proficiency training for engineering

Figure 4.5
DDG-51s Without an Embedded Engineering Training Capability, by Fiscal Year, Norfolk- and San Diego–Based Ships

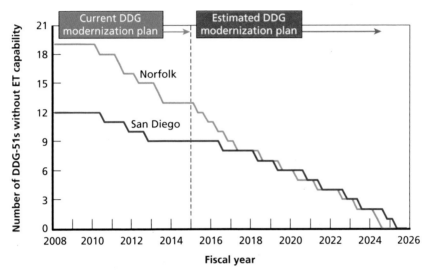

RAND *MG874-4.5*

evolutions and drills. Recently, Smartship cruisers with an OBT have been directed to conduct their ULTRA-Es and EOCs in port rather than underway, as traditionally done. These exercises are done with the engineering plant lit off and the main engines at idle. The onboard trainer is used as a casualty initiation tool, and the CCS watchstanders take ECC actions on the actual equipment just as if an actual casualty would occur. The drills must involve all subordinate watchstanders, and ECC actions must be completed through supplemental actions.

When drills are conducted this way, the CCS watchstander's actions are the same whether the casualty is imposed in port or underway. There are some limitations to a cruiser's drills with the OBT, including the inability to conduct some shafting drills in port. The onboard trainer can also be slow to operate, can give erroneous alarms, and offers no interaction with operating equipment. Nevertheless, the onboard trainer in cruisers does allow engineering training to take place without affecting ship movements or operations. Most drills can be done in port without any degradation of fidelity. The trainer offers an effective way to assess supervisory watchstander's adherence to procedures.

The Navy's use of OBTs in cruisers for conducting training ULTRA-Es and EOCs in port has direct implications for the newly installed and backfitted embedded trainers on DDG-51–class ships. Specifically, if the cruiser's OBT can be used to impose drills on the engineering team, then the DDG-51 embedded trainer can be used for the same purpose.

How long it will take to develop this ability on all DDG-51–class ships is unclear. The embedded trainer is being fielded to older DDG-51s at a slow rate. Without a change to policies and approaches to conducting engineering training, ships will continue to need a great number of repetition and underway days and other resources to conduct engineering training.

In the next chapter, we consider more generally the advantages and disadvantages of the DDG-51 shore-based engineering simulator.

CHAPTER FIVE

Pros and Cons of an Engineering Simulator

Potential Benefits and Shortcomings of an Engineering Simulator System

Training in an engineering simulator is not completely identical to training onboard the ship. Nevertheless, as we have seen and discuss further, there are many benefits of an engineering simulator.

Simulator training is not as susceptible as underway training is to cancellations or modifications because of weather, mechanical problems, other command tasking, or other conflicts. A simulator also does not involve extensive operations such as sea and anchor details, transit to and from an operating area (often 50 miles away from land) and ship maneuvering and control (safety of navigation and maintaining propulsion) issues. Simulators also do not require the resources (fuel, other consumables, wear and tear on the ship, and time of other crew members needed to get underway) that underway training does.

The ability to quickly reset the simulator to redo a drill or initiate a new one multiplies the amount of training that can be done, especially in comparison to that done underway. Unlike an underway drill, repeating a simulated drill does not require repositioning personnel or establishing communication and the time needed to do that. And because drill responses must be committed to memory, frequent and rapid repetition improves training.

The replay capability of the engineering simulator—reviewing casualty control actions from recorded data—also makes drill reconstructions and debriefings more efficient. Less time is needed to discuss *whether* something happened, leaving more time to discuss *how* and *why* things happened and what improvements could be made.

Engineering simulators also provide opportunities for types of training that are extremely difficult to accomplish underway. For example, the simulator systems can provide training to cope with casualty control, such as major system failures, and other casualties, that cannot be practiced realistically during live training underway.

The simulator can also complement and support more-basic underway training needs. For example, the engineering simulator can be used to practice a training event prior to conducting the same event onboard the ship, thereby increasing the efficiency of the available ship underway training time.

One drawback of the shore-based simulator is that engineers scheduled for simulator training are not available for shipboard responsibilities during the preparation, execution, and reconstruction of the engineering simulator training. The converse, of course, is also true, inasmuch as engineers scheduled for underway training are likewise unavailable for other ship responsibilities for extended periods during the engineering training period, but they are onboard and available if needed.

Upgrading junior engineers and those with little experience on console operations requires supervision from the ETT and/ or ATG instructors. Conducting training in a shore-based simulator for inexperienced engineers and those with lower qualification levels would require a reduced level of supervision as compared to underway training. During underway training, it is desirable that a watchstander who takes an action during a drill be supervised by a ETT member to ensure ship safety, because a watchstander's mistaken action could cause actual damage. With simulators, there is no concern for damage and fewer trainers may be employed. Moreover, simulator training ashore could provide more-standardized training across all ships, with well-trained instructors training watchstanders to a common standard. Still, it is difficult to simulate the psychological stress that can be caused by anticipation of the catastrophic consequences of serious errors or lapses in judgment. Highly qualified, experienced engineers can maintain their skills more reliably using simulator training, as can junior, less-experienced engineers.

Evaluation of Training Options—At Sea, Pierside, or Shore-Based

We compared and contrasted the conduct of engineering casualty control drill training as it is done on the ship at sea, on the ship pierside, or in a shore-based simulator. Table 5.1 presents this comparison in a stoplight format—green being good or best, yellow being neutral, and red being poor or least attractive—by variables such as cost, training constraints, and cohesiveness.

We do not weight these variables factors, but we understand that some—cost, for example—are more important than others. In this comparison, a shore-based simulator offers many advantages over training done onboard at sea or pierside in port. These include cost, cueing of watchstanders, trainee feedback, reduced energy use, and safety of training. A shore-based simulator offers an opportunity for conducting drills, and such a simulator could be part of a balanced approach to improving training.

Factors That Affect the Use and Acquisition of Simulators

In addition to comparing simulators with underway training, we also posit factors that might affect the use of simulation for training. Table 5.2 lists those associated with the characteristics of a simulator and alternative methods of performing training. We generalized to identify factors that would be important for increased use of simulation or would support or hinder it.

Among the most important factors for increased use of simulation are close location, high fidelity, a broad range of exercises that may be performed, flexible times for use, and a return on investment. By contrast, a removed distance, low or poor fidelity, a limited range of exercises that can be performed, and high cost all hinder or limit the acquisition and use of simulators for training.

Some of these factors are related. For example, the configuration of a simulator has a bearing on its operational realism. Some factors may combine to support or limit the use of simulation. Returns on

Table 5.1
Factors to Consider in Using Shore-Based Simulators or Shipboard Equipment for Training

Factor/Location of Training	At Sea	Pierside	Shore-Based Simulator
Cost	High fuel costs plus wear and tear	Lower cost, but wear and tear	Lower cost, no wear and tear
Operate own ship's equipment	All engineering equipment can be operated	Some can be operated, but not all	Ship's equipment not operated
Cueing of watchstanders	Some cueing by training team on drill imposition	Some cueing by training teams on drill imposition	No cueing
Number of ECC drills than can be done	All 40	32 of 40	35 of 40
Time available by crew for training	Dedicated crew underway, but underway time is decreasing	Maintenance demands in port are high. CCS is hub of activity in port—conflicts will arise	No conflicts, but competes with other unit's training needs
Training constraints	ECC drills normally done underway on a not-to-interfere basis with other training needs and/or impact bridge operations or ship's ability to navigate	Some conflicts with in-port maintenance demands and other ship events	Trainees must leave ship for training. Must trade off what they would be doing if they stayed on board, and what doesn't get done
Who gets trained	2 of 3 Engineering Watch Teams composed of CCS and in-space watchstanders	2 of 3 Engineering Watch Teams composed of CCS and in-space watchstanders	3 of 3 CCS watchstanders but not in-space watchstanders
Personnel involved in training	ETT and all watchstanders	ETT and all watchstanders	CCS personnel
Impact on nonengineering watchstanders	Electrical load limitations for combat systems, navigation and bridge equipment	Small impact	None
Usefulness to utilize for varying skill levels	Good for experienced and inexperienced personnel, but expensive and potentially hazardous if incorrect actions taken	Good for experienced and inexperienced personnel and less expensive; potentially hazardous if incorrect actions taken	Good for experienced and inexperienced personnel and least expensive over time; good for continuation training
Impact of watchstander errors	CCS personnel and in-space watchstanders – potential for being costly and dangerous	CCS personnel and in-space watchstanders— potential for being costly and dangerous	Trains CCS personnel only— no hazard to personnel or equipment

Table 5.1—Continued

Factor/Location of Training	At Sea	Pierside	Shore-Based Simulator
Feedback mechanism to trainees	In-space watchstanders stopped for safety violations. CCS watchstanders will perform immediate and controlling actions—graded as effective or ineffective based on observation and written comments about their actions	In-space watchstanders stopped for safety violations. CCS watchstanders will perform immediate and controlling actions—graded as effective or ineffective based on observation and written comments about their actions	Program can be "frozen" to provide instruction to watchstanders. Printout of time and sequence of actions offer ability to trace actions and timeline and provide objective feedback
Time it takes to conduct training	Longer. Must be approved by commanding officer and deconflicted with other training events onboard	Long. Deconfliction is required with ongoing maintenance and other shipside training	Short. Provides list of drills and runs training events. Events may be repeated to ensure proficiency
Maintenance of Engineering Training Team (ETT) Casualty Control Proficiency	Proficiency of ETT is unknown and untested	Proficiency of ETT is unknown and untested	Good. ETT members receive proficiency training as well as 1st and 2nd watch teams; ECC drill proficiency can be maintained in a shore-based simulator
Engineering watchstander's cohesion	Good. All are trained and communicate together	Good. All are trained and communicate together	Good for CCS watchstanders only
Physiological—heat, sound, sight, smell, ship movement, stresses	The real thing	Fewer stresses in port	Simulated environment
Realism of drill imposition	Some impositions different from an actual casualty, e.g., grease pencil used to indicate a high tank level	Simulations and deviations exist	Casualties alarm and occur to CCS watchstanders as they would underway
Effectiveness standard	Underway demonstration standard is 50%	Onboard demonstration standard is 50%	Can be trained to a higher effectiveness standard
Energy savings/carbon footprint	High energy use	Reduced energy use	Little energy use
Safety	Proficiency gained on operating equipment	Proficiency gained on operating equipment	Safe. Can train and gain proficiency before getting underway

Table 5.2
Factors That Could Affect Use of an Engineering Simulator

Simulator Factor	Important for Increased Use	Supports Increased Use	Hinders or Limits Increased Use
Physical location	Onboard the ship or close by	Close by; readily available	Not close or readily available
Fidelity	High	Medium	Low or poor
Operationally realistic	High; closely resembles operating conditions and/ or environment. Depicts near-actual scenarios	Nearly, but some differences	Little in common with operating conditions, environment. Does not adequately represent operation
Range/number of exercises that may be performed	Numerous, many or few, but important for readiness	Some, but not all	Few
Equipment	Closely resembles onboard equipment	Nearly resembles; some differences	Does not resemble
Training time available for use	Flexible	Available, but limited	Inflexible to unit's needs
Simulator configuration updates	Updated/upgraded as ships are updated	Generic, but close	Not updated/ upgraded over time
Training standards	Common		Different
Return on Investment	Provides a proven training benefit and/or savings	Helps training with little or no savings	Provides little training benefit and/or savings
Simulator cost	Training value exceeds costs	High or low cost; good training value	High cost; little training value
Risk	Reduces	Some reduction	Little reduction

investment and simulator costs are closely related and may alone support or not support increased use of simulation. Risk reduction is a critical factor that would support and drive increased use of a simulator. Overall, we posit that simulation should be pursued as a training alternative when it can sustain readiness, enhance a capability, save resources, or reduce risk.

How will simulators gain more widespread use? They must be realistic and present an adequate representation of events that would be encountered live. They also must be able to show increased performance—that is, they must save time or money or make crews more proficient and have an adequate return on investment.

We turn next to the policies and resources that could support simulator use.

CHAPTER SIX

Resourcing and Policy Changes Needed

Resourcing Needed

The DDG-51 desktop trainer software is available now and is already used at the Surface Warfare Officer School. The software was developed for Naval Air Warfare Command, Training Systems Division (NAWC TSD). Through our discussions with NAWC TSD officials, we learned that the software is government issue and therefore free for distribution to ships and shore training commands. There are some computer hardware requirements to fully use the software capability, but the software can be run on a shipboard computer.

The Center for Naval Engineering (CNE) has shore training sites at Norfolk and San Diego, as well as other locations. These sites have 12 dual-monitor computers that can run the DDG-51 desktop trainer. NAWC TSD officials have indicated they have plans to install the DDG-51 software at these CNE sites.

The DDG-51 engineering console trainer, like the one used at SWOS, costs $1.6 million per console to procure and $300,000 to install. Sustainment costs include an operator, a technician, and software updates as needed.

An engineering simulator could produce fuel savings and increased readiness. Fuel prices have varied over time, and future fuel prices are uncertain. Figure 6.1 shows the tradeoff between the price of oil per

Figure 6.1
**Estimated Underway Days That Must Be Saved Over a Ten-Year Period to
Offset Acquisition and Sustainment Costs of a DDG-51 Simulator**

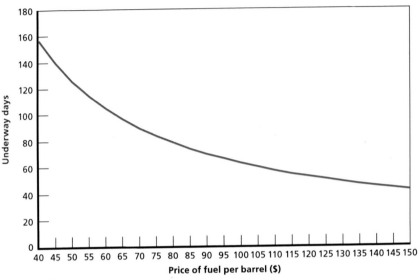

RAND MG874-6.1

barrel[1] and the number of underway days that must be saved over a ten-year period to recoup the acquisition and sustainment costs for a single engineering simulator.

Should fuel cost $40 per barrel, a level it was approaching in late 2008, a simulator would have to eliminate the need for nearly 160 underway training days over a decade to offset its acquisition and sustainment costs. Conversely, if fuel should cost $150 per barrel, a level it approached earlier in 2008, a simulator would have to eliminate the need for only about 40 days of underway training in a decade, or only about four days per year, to offset acquisition and sustainment costs. We cannot say with precision how other potential savings (wear and tear, maintenance, and labor) would affect this tradeoff. However, cost savings can certainly be achieved because the starting, operation,

---

[1] The figure reflects the *unburdened* cost of fuel. If the burdened cost of fuel were considered, which includes fuel transportation costs, the number of underway days needed to offset acquisition and sustainment costs would be reduced.

and stopping of equipment during each underway day increases wear and tear—which, in turn, increases the demand for maintenance and labor.

We estimate that approximately 50 Norfolk-based ships and 39 San Diego–based ships (without an embedded training capability) will undergo ULT from FY 2009 to FY 2018. Even if fuel were just $40 per barrel over the next decade, an engineering simulator in Norfolk would pay for itself if it saved only about three underway days per ship over the decade, while a simulator in San Diego would pay for itself if it saved only about four days per ship.

Policy Changes Needed

Policy changes can and should be made to incorporate engineering simulator training into the training mix. Above all, if the Navy were to purchase a DDG-51 engineering console, it should make DDG-51 simulator use part of the training process. This would require ship's engineers to meet or exceed engineering proficiency standards before going to sea. Policies are necessary for granting credit for continuation or repetitive training to meet proficiency standards on an engineering console. Ships should also be required to use the engineering simulator for training during extended yard periods. Training policies should stipulate that the drills and proficiency of the watchstanders be documented and that underway training be tailored to meet those proficiencies that cannot be reached or evaluated on an engineering simulator. SWOS should be consulted about placing console trainers at Norfolk and San Diego and should provide direction on how the simulator can and should be used to support PQS qualifications for prospective CCS watchstanders. Scheduled simulator training for senior enlisted personnel during their en-route PCS pipeline to their ships should also be required.

The desktop trainer is a valuable training resource and should be used onboard ship for console training for the seven consoles that it simulates. Current and prospective watchstanders should use the desktop trainer for initial console familiarization and training and to prepare for advanced console operations.

The desktop trainer can also be used at shore facilities, such as at CNE sites. With little or no cost, the software could be loaded onto CNE classroom computers via a local area network (onboard ships or ashore) for student training. A training course should be developed for operators, similar to the class of instruction at SWOS, to exploit this training resource to its fullest extent. Although ships at Norfolk and San Diego would have their own console trainers, such a course would be particularly important for ships located at homeports without a shore-based console trainer. A training progression should be developed for CCS watchstanders, so that they may train or gain proficiency on the desktop trainer prior to underway operations.

CHAPTER SEVEN

Findings and Observations

Much of the training to improve engineering evolution and casualty control drill proficiency is presently done underway. However, our research has shown that all evolutions can be done in port and that many casualty control drills can be done in port as well.

Underway training is costly. Ships burn large quantities of fuel and incur equipment wear and tear that increases maintenance demands. The use of a shore-based engineering simulator console could improve watchstanders' proficiency, reduce ULT underway training days—saving fuel and reducing equipment wear and tear—and potentially result in the ship being ready for tasking earlier in the training cycle.

Although some ships meet proficiency standards at the start of training, most ships need ATG assistance and conduct a great deal of training underway to improve their proficiency. Evolution and casualty control drill proficiency gradually improves over time. Most ships meet evolution proficiency standards after a short period of training and tend to sustain that proficiency, but proficiency standards for casualty control drills are harder to achieve.

Our interviews and ship visits indicate that CCS is a hub of activity in port. While ships are in port, the engineering department focuses on the coordination of maintenance, troubleshooting, and repairs, not on training, making it challenging to train in port (and underway) without a simulator.

Subject matter experts from the ATGs told us that much manpower is dedicated to improving engineering readiness on surface combatants. Our data indicated that for Atlantic-based ships, five to thirteen TG engineers get underway with ships to conduct training and/ or assessments.

Other navies use simulators for engineering qualifications and continuation training to a greater degree than U.S. Navy does. But the engineering departments of other navies are also organized and trained differently. Those navies have professional engineering officers and employ a smaller number of watchstanders, resulting in lower training demands.

Embedded training is now standard on all ships with the most modern control systems. The organizations we studied do not yet appear to have considered how to balance embedded training, shore simulators, and at-sea training. Of the organizations we examined, the British Royal Navy and the Royal New Zealand Navy appear to have considered these balance issues the most.

The Navy already has a capable training resource in both the engineering desktop trainer and the console trainer. These trainers offer a means to improve CCS watchstander proficiency and could produce many benefits. The benefits include having *all* CCS watchstanders receive training—not just two of the three watch teams, as is currently done. The trainers can be used for EOC preparations as well as for continuation and refresher training. They provide a great capability to quickly restore crew proficiency after an extended maintenance period. A simulator is also a more efficient and safer way to train.

An engineering simulator offers the opportunity for CCS engineering watchstanders to meet engineering casualty control standards before going to sea. If the Navy were to place simulators in Norfolk and San Diego, more enlisted personnel—those who actually stand the watch—would have an opportunity to train the way that officers currently do in Newport. Simulators could also allow personnel to complete PQS qualifications. Increased use of a simulator may also reduce demands on ATG engineers by allowing ship personnel to achieve a higher level of proficiency before undergoing ATG-assisted training.

An investment in engineering simulators is needed. The desktop console software is available now, with little cost to install. The software belongs to the government, and the program can be run in stand-alone mode or networked. Some computer hardware requirements must be met to run the software.

The DDG-51 engineering console trainer costs $1.6 million for the initial outlay and $300,000 for one-time setup costs and sustainment costs. The same government-issue software that runs the desktop trainer also runs the console trainer. The major cost of the initial outlay is that to build the console hardware panels. The sustainment costs include an E-6 operator, a technician, and software updates as required. The government can realize a return on this investment by reducing underway days for training and their associated costs (e.g., fuel).

Steps to Take Now

The Navy DDG-51 class should distribute DDG-51 engineering desktop trainer software to ships and to ATGLANT and ATGPAC now. This software program should be used by engineering watchstanders onboard their own ships or ashore in classrooms. The desktop trainer could be a low-cost option to increase engineering watchstander proficiency for ships that are not located at Norfolk or San Diego.

The Navy could install DDG-51 console trainers at Norfolk and San Diego. Desktop trainer and console trainer use could be modeled after the class of instruction at SWOS.

The Navy is planning to backfit DDG-51s with an embedded training capability. Accelerating the installation of embedded DDG-51 engineering trainers could allow for an improved method of training. Embedded engineering trainers allow training to be done on a ship's own equipment, which is perhaps the best training approach. Nevertheless, we realize that this may be a costly and time-consuming option.

Policy Changes Needed

If the Navy were to proceed with installation plans for the engineering desktop trainer and console trainer, it would need to make DDG-51 simulator use part of the training process. Ship engineering watch teams should have to qualify on the shoreside console simulator before getting underway. If the simulator were not made part of the process, it might not be used enough to justify the investment it requires. Policy changes needed to increase engineering training proficiency through simulators should include the following:

1. Use the desktop trainer as a lead-in trainer for advanced operations.
2. Perform alignment, starting/stopping, and MLOC plant operations on the desktop trainer.
3. Install the desktop trainer software onboard ships and ashore. Provide the software to ships located at homeports without a console trainer, and at Norfolk and San Diego, prior to a console trainer installation.
4. For ships whose homeports have console trainers, CCS watchstanders should train and gain proficiency on the console trainer in ECC drills. CCS watchstanders should be required to meet or exceed the fleet proficiency standard in the console trainer before going to sea.
5. Use the console trainer during extended yard periods and for repetitive training requirements.
6. SWOS should evaluate fleet use of the shore-based engineering console simulators identical to those at Newport, and should provide recommendations and authority for use in support of PQS qualifications.

List of EOCC Drills and Evolutions Required for DDG-51–Class Ships

This appendix shows the EOCC drills and evolutions required for engineering proficiency training for DDG-51–class ships. Table A.1 lists the drills, and Table A.2 lists evolutions.

Table A.1
EOCC Drills

Drills	Validations
Main Engine Drill Family–Category 1 Drills (quarterly, core)	
MMFOL—major fuel oil leak	None
MBGTM—class Bravo fire in GTM	None
Main Engine Drill Family–Category 2 Drills (semiannually, elective)	
MLFOP—loss of fuel oil pressure	None
MGGS—gas generator stall in GTM	MLPTO, MEPTV, MGGOSMHTIT, MLPLAMPTOS
MECUF—executive control unit failure	None
MLPACC—loss of propulsion and auxiliary control console	MLSCU
MLSCU—loss of shaft control unit	MLPACC
Main Engine Drill Family—Category 3 Drills (annually)	
MCASF—gas turbine cool air system failure	None

Table A.1—Continued

Drills	Validations
MLPTO—low lube-oil pressure in GTM	MGGS, MEPTV, MGGOSMHTIT, MLPLAMPTOS
MEPTV—power turbine vibrations high in GTM	MGGS, MLPTO, MGGOSMHTIT, MLPLAMPTOS
MGGOS—gas generator overspeed in GTM	MGGS, MLPTO, MEPTVMHTIT, MLPLAMPTOS
MHTIT—power turbine inlet temperature high in GTM	MGGS, MLPTO, MEPTVMGGOS, MLPLAMPTOS
MLPLA—loss of PLA in GTM	MGGS, MLPTO, MEPTVMGGOS, MHTITMPTOS
MPTOS—power turbine overspeed in GTM	MGGS, MLPTO, MEPTVMGGOS, MHTITMLPLA
MPSFP—post-shutdown fire in GTM	None
Propulsion Drive Train Family–Category 1 Drills (quarterly, core)	
MLLOPR—loss of lube-oil pressure in main reduction gear	None
MHBRG—hot bearing red gear	MHLSB, MHST, MNVRG
Propulsion Drive Train Family—Category 2 Drills (semiannually, elective)	
MLCRP—loss of pitch control	None
MLHOL—major leak of CRP/CPP system	MLLOL
MLLOL—major lube oil leak in main reduction gear	MLHOL
MNVRG—noise/vibration in main reduction gear/shaft	HBRG, MHST, HLSB
MHLSB—hot line shaft bearing	MHBRG, MHST, MNVRG
MLHOP—loss of CRP/CPP pressure	None
Propulsion Drive Train Family—Category 3 Drills (annually)	
MHST—high shaft torque	None

Table A.1—Continued

Drills	Validations
Electrical Family–Category 1 Drills (quarterly, core)	
MOSGG—overspeed GTG	None
MBGGM—class Bravo fire in GTG module	None
Electrical Family—Category 2 Drills (semiannually, elective)	
MHBGTG—hot bearing GTG	MNVGG, MGHIT, MLGGO
MGHIT—high gas turbine inlet temperature GTG	MHBGTG, MNVG, MLGGO (DDG only)
MLGGO—loss lube oil pressure GTG	MHBGTG, MNVG, MHIT (DDG only)
MPSFG—post-shutdown fire GTG	MPSFR
MPSFR—post-shutdown fire in engine	MPSFG
MCCFG—class Charlie fire generator	None
Electrical Family—Category 3 Drills (annually)	
MLEPC—loss of EPCC (MLMCS—loss of control console in smart ship)	None
MFZDB—electrical fault on zonal main bus	None
MNVGG—unusual noise/vibration in GTG	MHBGTG, MGHIT, MLGGO (DDG only)
Integrated Family—Category 1 Drills (quarterly, core)	
MCBF—class Bravo fire in main space	None
MCFED—class Charlie fire in electrical distribution system	None
MMF—flooding in main space	None
Integrated Family—Category 2 Drills (semiannually, elective)	
MCCFS—class Charlie fire in switchboard	None
MLSC—loss of steering control	None
Integrated Family—Category 3 Drills (annually)	
MLCWS—loss of chill water	None

The engineering evolutions shown in Table A.2 must be performed by the designated DDG-51–class engineering watchstanders at the specified intervals.

Table A.2
Engineering Evolutions

Evolution	Frequency
EOOW routine evolutions	Quarterly, core
Don EEBD	
Evaluate heat stress survey	
Evaluate tag-out sheet	
Evaluate lube oil sample	
Evaluate fuel oil sample	
Review operating logs	
Start/stop fire pump	
PACC/PCC routine evolutions	Quarterly, core
Don EEBD	
Transfer control between PACC/PCC, PLCC SCU, and LOP	
Transfer control between PACC/PCC and SCC	
Shift fuel oil pumps	
Shift lube oil pumps	
Motor GTM	
Start GTM	
Stop GTM	
Test console alarms	
Start/stop sea water service pump	
EPCC routine evolutions	Quarterly, core
Don EEBD	
Start/parallel GTG/SSDG	
Parallel bus to bus	
Remove load/stop GTG/SSDG	
Test EPCC alarms	
Engine room routine evolutions	Quarterly, core
Don EEBD	
Align main reduction gear lube oil cooler	
Shift lube oil strainers/filters	
Shift purifier suction	
Align/operate/secure eductor	
Verify/align GTG for standby	
Draw lube oil/CPP/purifier efficiency sample	
Draw lube oil cooler waterside sample	
Shift low pressure air compressor mode	
Verify alignment stern tube cooling	
Align/start/operate/secure oily waste transfer pump	

Table A.2—Continued

Evolution	Frequency
Align/start HPAC	
Align/start L/O purifier	
Test SCU/PLCC/PLC alarms and indications	
Engine room infrequent evolutions	Annually, elective
Verify/align GTM fuel oil system	
Verify/align GTG support systems	
Verify/align fire pump	
Shift lube oil pumps	
Start GTM	
Stop GTM	
Motor GTM	
Start GTG	
Motor GTG	
Shift fuel oil service pumps	
Fuel purge GTM	
Start/stop fire pump	
Start/stop sea water service pump	
Align/secure anti-icing	
Auxiliary equipment routine evolutions	Quarterly, core
Don EEBD	
Align/operate/secure eductor	
Draw lube oil sample	
Align/start evaporator	
Align/start high pressure air compressor	
Sample/test potable water	
Align freshwater tank to fill	
Verify alignment stern tube cooling	
Align/start/stop air conditioning plant	
Auxiliary equipment infrequent evolutions	Annually, elective
Start/stop fire pump	
Verify/align fire pump	
Start/stop sea water service pump	
Oil lab routine evolutions	Quarterly, core
Don EEBD	
Draw coalescer outlet/bottom sample	
Draw prior to start sample on service/auxiliary service tank	
Conduct contaminated fuel detector/free water detector	
Align/operate/secure oily water separator	
Transfer fuel oil storage to service	
Recirculate fuel oil service tank	
Conduct auxiliary fuel oil transfer	

Table A.2—Continued

Evolution	Frequency
Oil lab infrequent evolutions	Annually, elective
Conduct lube oil BS&W Conduct fuel oil BS&W	
Sounding and security routine evolutions	Quarterly, core
Don EEBD Align/operate/secure eductor	
Sounding and security infrequent evolutions	Annually, elective
Verify/align fire pump	
Switchboard routine evolutions	Quarterly, core
Don EEBD Shift control to switchboard	
Switchboard infrequent evolutions	Annually, elective

Start/parallel GTG/SSDG
   Remove load/stop GTG/SSDG
   Parallel bus to bus	

MLOC evolutions	Quarterly, core

Test EOT
   Propeller pitch control test
   Verify/align bleed air system (including prairie and masker air)
   Inspect GTM module
   Verify/align GTM synthetic L/O
   Verify/align CRP/CPP system
   Inspect GTG module
   Verify/align LPAC
   Verify/align LPAD
   Verify/align MRG L/O system
   Engage/disengage turning gear
   Start turning gear forward/reverse direction
   Verify S/W cooling/service alignment
   Pressure/test L/O strainer
   Verify/align SWS pump
   Verify/align SWS system
   Verify/align F/O compensating system
   Verify/align F/O service system
   Start F/O pumps
   Start L/O pumps
   Start CRP/CPP pump
   Verify/align/test steering gear
   Verify/align AFFF system
   Shift to ship's power	

Engineering Equipment Contained in DDG-51 Engineering Spaces

DDG-51 Engineering Machinery Arrangement

There are seven main engineering spaces in a DDG-51 class destroyer. We list these below. Engineering personnel stand watch in CCS, the main engine rooms, and auxiliary spaces. CCS is located above Main Engine Room No. 2 and contains the consoles for operating and monitoring the engineering plant. A general description of a DDG-51 engineering plant is provided below.

The propulsion system is made up of two independent propulsion plants; each plant is self-sufficient so that malfunction of one will not affect the operation of the other. The propulsion plant consists of four gas turbine engines arranged in pairs. Each pair of turbines is directly connected to a reduction gear that has clutches and brakes to control the torque transmitted to the main shaft and the controllable pitch propeller.

The propulsion machinery in the forward engine room drives the starboard shaft and the machinery in the aft engine room drives the port shaft. Each controllable-pitch propeller has five blades. Each propulsion gas turbine is located within a module and is provided with its own lubrication system, electronics, fuel and speed governing system, fire protection system, and control instrumentation from external support equipment.

Propulsion control is accomplished via the MCS. Propulsion controls and displays are provided at major consoles located in the pilot house, CCS and the two engine rooms. Lubricating oil is supplied to the main reduction gear and the propulsion gas turbine modules. The propulsion fuel service system supplies fuel to the propulsion gas turbines.

Engineers need to know the plant interactions and effects of equipment casualties on other equipment. The equipment and machinery located in each engineering space are detailed below:

a. Forward Vacuum Collection and Holding Tank (VCHT) contains Sewage Treatment Plant Number 1 and Number 1 fire pump

b. AUX 1 contains the following major equipment:
   - Number 1 Gas Turbine Generator (GTG)—sAllison 501-K34 (2500KW, 450V)
   - Number 1 Ship Service Switchboard
   - Number 2 Fire pump
   - Number 1 Seawater Cooling Pump
   - Number 1 A/C Plant
   - Number 1 and Number 2 Potable Water Pumps
   - A fresh water fire fighting pump

c. MER 1 contains the following major equipment:
   - 1A and 1B Gas Turbine Modules
   - Number 1 Main Reduction Gear (MRG)
   - 1 and 2 Vapor Compression Distillers
   - Number 3 Fire pump
   - Number 2 Seawater Cooling Pump
   - Number 1 HP Air Compressor
   - Number 1 LP Air Compressor
   - Number 1 Fuel Oil Transfer and Purification System
   - Number 1 Fuel Oil Service System
   - Number 1 Lube Oil (L/O) Service system
   - Number 1 Controllable Reversible Pitch (CRP) System

d. AUX 2
   - Number 2 and Number 3 A/C Plants
   - Number 3 Seawater Cooling Pump
   - Main Thrust Bearing Starboard Shaft

- Oily Water Separator
e. MER 2 contains the following major equipment:
  - 2A and 2B Gas Turbine Modules
  - Number 2 Main Reduction Gear
  - Number 2 GTG
  - Number 2 Ship Service Switchboard
  - Number 4 Fire pump
  - Number 4 Seawater Cooling Pump
  - Number 2 HP Air Compressor
  - Number 2 and Number 3 LP Air Compressors
  - 1A Line Shaft Bearing
  - Number 2 Fuel Oil Transfer and Purification System
  - Number 2 Fuel Oil Service System
  - Number 2 L/O Service system
  - Number 2 CRP System
f. A/C Machinery/Pump room (Shaft Alley)
  - Number 4 A/C Plant
  - Number 5 Fire pump
  - Number 5 Seawater Cooling Pump
  - 1B and 2A Line Shaft Bearings
  - Port Shaft Thrust Bearing
g. Generator Room
  - Number 3 GTG
  - Number 3 Ship Service Switchboard
  - Number 6 Fire pump
  - JP-5 Fill and Transfer System
h. After steering
  - Number 1A and 1B Hydraulic Power Units (HPU)
  - Number 2A and 2B Hydraulic Power Units (HPU)
i. After Steering Unit for steering from aft steering

How Commercial Industry and Other Navies Use Simulators for Engineering Training

Previous RAND work has investigated more broadly how other organizations use simulation to assist in training.[1] In this project we researched the commercial industry and other navies to understand better how they use simulators for engineering training. Specifically, we sought to understand the following:

- How do these other organizations man their seagoing engineering watches?
- Who are the watchstanders?
- How are they organized?
- How are individual watchstanders and watch teams trained?
- What are the pre-joining or educational training requirements?
- How is training accomplished in shore schools?
- How is training accomplished underway?
- How are simulators used?
- What types of simulators are available?
- How are they used?

We researched and interviewed representatives from five organizations: the U.S. Merchant Marine, the U.S. Coast Guard, the British Royal Navy, the Canadian Navy, and the Royal New Zealand Navy. From this research and subsequent analysis, we are able to describe the

---

[1] John F. Schank, Harry J. Thie, Clifford M. Graf, II, Joseph Beel, and Jerry M. Sollinger, *Finding the Right Balance: Simulator and Live Training for Navy Units*, Santa Monica, Calif.: RAND Corporation, MR-1441-NAVY, 2002.

similarities and differences of engineering simulator training between these other organizations and that for DDG-51 vessels.

We asked each organization to describe how it mans the engineering departments of its ships, the training it gives to watchstanders before joining and on a sea-going watch, and how it uses engineering simulators in their training programs or even to replace some underway training time.

To best make comparisons with DDG-51 training, we use U.S. Navy terminology where possible. For example, all of the navies use senior enlisted personnel as Engineering Officers of the Watch, but they each have different rank structures and nomenclature. Except when we wish to highlight substantive differences, we use U.S. Navy terminology to describe positions and their responsibilities, noting in parentheses the actual terminology other organizations use.

U.S. Merchant Marine

The U.S. Merchant Marine—that is, U.S. flagged ships—has for some time been reducing its manning to lower running costs. At the same time, it has optimized its preparatory and ship-specific training. A commercial ship is very different from a Navy surface combatant. It has a simple routine of loading cargo, moving to a new port, unloading the cargo, and then repeating the process. There is no time at sea for delaying drills or crew training; this would increase costs and disrupt delivery schedules.

Merchant ships are designed for efficient and cost-effective operation with high levels of automation and unmanned spaces. They have moved from manning in the engine spaces, which was necessary when the older steam plants were in use, to enclosed operating stations that are air conditioned, often with windows overlooking the engine room or completely remote from the engineering spaces.

In particular, the change to diesel engines has made it easier to use remote monitoring, with the engine department on more modern ships working normal days with a duty engineer overnight. The overnight duty engineer has all the systems information available in his cabin and passes propulsion control to the bridge. The engineering department works normal days and does the maintenance or checks that are needed. Consequently, modern merchant ships have very small crews regardless of vessel size. For example, the Maersk line has a class of container ships[2] each of which weighs more than 170,000 gross tons, carries more than 11,000 containers, and has a crew of 13, of which four are EOOWs.

The U.S. Coast Guard sets the minimum acceptable manning for U.S. flagged ships in accordance with international and federal requirements. Each vessel has four EOOWs who are officers who have graduated from a maritime academy. Each also has four enlisted watch-standers known as Oilers, who have all the roving responsibilities of the external roundsmen, such as the oil king, S&S personnel, and space watchstanders in the Navy. A watch consists of the officer EOOW and an Oiler. The Chief Engineer leads the engineering department.

The vast majority of engineering training is undertaken prior to joining a ship. Trainees will achieve a graduate qualification at one of seven maritime academies (six state and one federal) and leave as 3rd Mates able to join a ship and keep a watch straightaway. In most cases, however, a Chief Engineer will arrange for new officers to undertake on-the-job training to become familiar with specific systems of the ship. Some shipping lines require additional specialist training courses (for ships with specialized equipment) at schools such as the Calhoon MEBA Engineering School.[3]

The U.S. Coast Guard National Maritime Center oversees the curricula at the academies and requires the undergraduate courses to include 180 days at sea in a training ship before the award of an engineer license. There are no training ships at the schools that deliver the specialist training. Rather, those schools rely on practical training with adapted "real" systems, for example, specially configured refrigerated containers or simulators that can provide sufficient training to meet the course standards.

---

[2] *Emma Maersk*–Class—8 ships in the class.

[3] MEBA is the Marine Engineers Beneficial Association.

Simulators are not used to provide a shore-based like-for-like alternative to the seagoing systems that will be used in the ships. For the most part, this is because there is little standardization across merchant ships, even those of the same line, so a school's investment in any one specific simulator would likely not be useful. Instead, the emphasis is on refining the engineering principles taught at the academies. This approach is reinforced by the requirements of the Coast Guard.

U.S. Coast Guard

Most Coast Guard cutters are much smaller than a DDG-51 and face a different set of broad demands placed upon them. To some extent, crews receive one month of training by U.S. Navy ATG engineers. Cutters operate to some extent similar to a merchant ship: vessels are at sea on transit for life-saving or inspection duties, for example, executing those duties, or returning to port. Port time is spent meeting maintenance requirements exacerbated by the lack of at-sea flexibility to work on systems that are needed for the mission.

The larger cutters have a watch organization very similar to that of DDG-51 ships, including an EOOW, PACC, EPCC and oil king. The oil king, however, has slightly modified, greater responsibility as the engine spaces are unmanned. Except where an engineer officer is standing a watch for career training, the watch comprises only enlisted personnel. The watches rotate in the same way as they do on a Navy vessel.

Individual watchstanders receive training broadly comparable to that of Navy counterparts with PQS and other similar pre-joining requirements. This training is undertaken at pipeline schools that also provide additional or specialist training if needed by personnel joining specific cutters. Administrative instructions, nominally from the Commandant of the Coast Guard, specify the onboard training requirements for personnel. The commanding officer, advised by shore-based staff if necessary, authorizes personnel to stand watch.

If the operational requirements allow, a cutter will undertake certain types of underway training once per week. Cutters also have a schedule of mandatory annual training. The closest comparison to the Navy training process comes when a cutter is evaluated biennially by the Navy ATG.

Seagoing personnel do not use engineering simulators for training, and simulators are only used to a limited extent in the pipeline training schools. This is likely to change as the new National Security Cutter enters service. To support these ships, new simulators are being built and these may offer new ways of training personnel.

Royal Navy

The British Royal Navy is introducing a new class of warship, the Type 45 (T45) *Daring* class, which is very similar in role and size to the U.S. Navy's DDG-51 class. The T45s take advantage of the latest advances in engineering technology and will build on established manning and training processes.

The Royal Navy uses enlisted personnel for its engineering watches. One watchstander handles the combined duties of an EOOW, PACC and EPCC. Oil king duties are handled by one watchstander on T45 ships and two watchstanders on other ships. Given unmanned engine spaces, Royal Navy oil kings have greater responsibility than their U.S. Navy counterparts have. Engineering officers will stand watch as part of their career training and, in such cases, supplement the established enlisted manning.

Engineering personnel receive extensive initial training followed by vessel or system-specific training prior to joining their first ship. At-sea personnel complete formal, documented on-the-job training before sitting for exams that, with the commanding officer's authorization, allow unsupervised watchstanding. Training continues at sea for qualified personnel as a prerequisite for professional advancement (for rank and pay) and subsequent follow-on career training at the engineering school.

Whole-watch casualty drill training is undertaken at least twice weekly for two hours per session. These drills are for engineering casualties affecting the position and intended movement of the ship and often culminate in drills for whole-ship emergencies that can involve damage control or fire-fighting teams. Other additional drills are completed within the watch and without external involvement. The two-hour drill sessions will often be split across a watch change to allow two teams to train.

Simulators are used extensively in the training schools for initial and career training. This training includes basic professional skills, as well as team training to familiarize personnel with seagoing drills. The Royal Navy has fully functioning CCS simulators for each major class of warship. These are used by ship's teams for proficiency training during pierside periods. Such training can meet the weekly training requirement.

At least once yearly a ship will undergo evaluation by the Flag Officer Sea Training organization. Harbor and underway engineering drills are taught first and then assessed to raise proficiency within each watch and across all watches and other sections of the ship. This training does not rely on the shore simulators.

The T45 engineering control system is based on the latest advances in system controls and will introduce new ways of training. The system does not rely on dedicated engineering terminals, although these are present in the CCS equivalent where the two watchstanders and roundsman will keep their watch. Rather, it uses "soft" touch-screen panels that are fully flexible. Any position within the CCS or elsewhere in the ship, such as the bridge or the Engineering Department Head's stateroom, can be configured to monitor and control any engineering system. Additionally, a panel or series of panels can be isolated to provide a synthetic training environment nominally using the ship's systems. The first T45 is completing initial sea trials and will enter service shortly.

Canadian Navy

The ships of the Canadian Navy are smaller than those of the DDG-51 class but they deploy worldwide and have similar advanced engineering systems. The most modern ships are those of the *Halifax* class. These have seen incremental upgrades to engineering systems since their entry into service.

Enlisted personnel undertake engineering watchstanding duties with officers responsible for the management of the department and professional engineering advice to the commanding officer. All personnel receive extensive initial training and specialist training prior to joining a ship. On-the-job training continues onboard until the new joiner is assessed as proficient to stand watch unsupervised. An engineering watch comprises an EOOW, EPCC, PACC, and two oil kings as roundsmen; machinery spaces are unmanned. Personnel are formally boarded at the schools to qualify in each rank and for the key watch positions of EOOW, EPCC, and PACC. Casualty drills are undertaken every day underway between 0700 and 0830 and can include whole-ship emergencies. Training occurs pierside using the CCS consoles, which are fully flexible and can be assigned any role, even at sea, where synthetic training on watch can take place.

Simulators are used in the fleet schools with an *Iroquois* Class CCS on the West Coast and a *Halifax* Class CCS on the East Coast. These simulators are exactly the same as the CCS spaces in the ships. They are used by career training courses and to refamiliarize ship personnel with specific drills or equipment controls.

Royal New Zealand Navy

Although the smallest organization of this group, the Royal New Zealand Navy is of interest because it has reviewed completely its training policy in anticipation of delivery of two new *ANZAC* frigates. These modern ships, though smaller than the DDG-51 Class, will have advanced engineering and control systems. These frigates together with other vessels will provide a regional capability and be able to deploy extended distances alongside allied nations.

The Royal New Zealand Navy training review was based on research that examined every aspect of training personnel, including pre-joining training, the training of personnel in ships individually and collectively, and career progression training. The Royal New Zealand

Navy found a Synthetic Training Environment (STE), generated by tailored training courses and comprehensive simulators could provide significant savings, reducing the time needed to train personnel and resulting in less wear and tear on the ships.

The engineering watch in ANZAC frigates will consist of three enlisted personnel in the CCS. One watchstander will handle EOOW, PACC and EPCC duties, while another will manage oil king duties and monitor unmanned machinery spaces. Engineering personnel will complete all their training before joining the ship with only a consolidation and endorsement period at sea on the actual equipment. This will allow the commanding officer and Engineer Officer to have confidence in the ability of each watchstander to undertake their responsibilities under all conditions. The watchstanders will be required to complete engineering casualty drills underway but will learn technical and leadership skills in the STE. The STE is expected to provide training for both normal and extraordinary operating conditions without endangering ship personnel or causing catastrophic and expensive failures of actual equipment. The STE will consist of full mission simulators enhanced by a generic Integrated Propulsion Management System trainer that will train main systems, auxiliaries, electrical and ship specific systems.

The STE is being acquired with the ANZAC frigates that are expected to enter service in 2009. It will likely take some time before any one ship will have personnel trained exclusively on the STE.

Comparing These Organizations with the U.S. Navy

Table C.1 summarizes engineering training in the above five organizations and on the DDG-51. While the unique global roles and responsibilities of the U.S. Navy make direct comparisons with other organizations difficult, there are a number of interesting similarities among the other organizations.

All the other organizations we reviewed operate their ships with unmanned machinery spaces. In the most extreme case, that of the U.S. Merchant Marine, it is feasible in the most modern ships for

Table C.1
Summary of Different Approaches to Engineering Training

		Organization					
		U.S. Merchant Marine	U.S. Coast Guard	USN DDG-51	Royal Navy	Canadian Navy	New Zealand Navy
Factors	Watch manning — Number of watch-standers	2	3 or 4	8	3 or 4	5	3
	Watch manning — Position/rank	EOOW (officer), oiler (enlisted)	EOW (CPO), throttleman (PO), generator (PO), oiler (enlisted)	EOOW, PACC, EPCC, ERO (2), ASM, EEM, PSM, oil king	MEOOW1 (CPO/PO), MEOOW2 (PO), 1 or 2 roundsmen (enlisted)	Chief of the Watch (PO), Machinery Control Operator (PO), Electrician of the Watch (PO), 2 roundsmen (enlisted)	EOOW (PO), machinery control operator (PO), roundsman (enlisted)
	Policy of team training ashore	None	None	Onboard. Can do drills in port or at sea. No simulator equivalency	Simulator equivalency places high demand on available systems that are located at only one base port	Shore training in CCS simulators optional	Ashore training integral to sea training and considered more cost-effective
	Policy of team training afloat	None. Reaction only to plant casualties	Similar to commercial. React to plant casualties	Surface training manual defines periodicy requirements; ships conduct drills at CO's discretion	Minimum 2x2 hours delay-ing drills per week with non-delaying and addi-tional drills when program allows. Drills roughly comparable to DDG51	Every day underway from 0700–0830. Drills roughly comparable to DDG51	That necessary for CO/EO confidence in personnel and team ability to operate machinery safely at sea
	Where and how simulators are used	Used ashore in training academies as part of education and qualification	None	Only in pre-joining training	When ship confined alongside, simulator equivalency for drills and w/k currency	Fleet schools and available for ship team use. One trainer west coast, another east coast	Synthetic Training Environment for all training
	Embedded training	None or limited	None	Newest DDG-51s have embedded engineering training capability; being backfitted into older DDG-51s	On latest ships using fully flexible soft panels	Partial	Will be determined as new class of ships introduced
	Team certification	Individual quals only	30 days with USN ATG every 2 years	Tailored engineering training; most done underway	Safety and readiness checks following maintenance periods, prior to sea training, 4 weeks sea training and pre-deployment	Process of at sea training and inspection following maintenance periods and prior to deployments	Bulk ashore in STE with validation and endorsement at sea

the engineering department to work only during the day and monitor the systems remotely from the bridge at night. The required size of the engineering watch is greatly reduced. More generally, more-modern ships have fewer watchstanders.

The individual training of engineering watchstanders varies across organizations. The greatest similarity in training appears to be among the U.S. Navy, the Canadian Navy, the Royal Navy, and the Royal New Zealand Navy. All use enlisted personnel as the watchstanders and a PQS equivalent system, undertake pre-joining training, and—to varying degrees—require on-the-job training or validation.

The underway training requirements of the navies are also broadly similar and very different from those of the U.S. Merchant Marine and the U.S. Coast Guard. All have similar casualty drills that vary by a ship's machinery. All have mandated underway training requirements for watchstanders.

The ability to undertake engineering watch training ashore is linked to the availability of suitable training resources, which tend to be CCS simulators. The most ambitious organization regarding watch training ashore is the Royal New Zealand Navy, which intends to establish a simulator-driven Synthetic Training Environment for all shore training. The Royal Navy and Canadian Navy have similar approaches to shore training, with the Royal Navy specifically allowing shore training in the appropriate simulators to be equivalent to the seagoing requirement.

Embedded training is now standard in all ships with the most-modern control systems. The organizations we studied do not yet appear to have considered how to balance embedded training, shore simulators, and at-sea training. Among the other organizations we examined, the Royal Navy and Royal New Zealand Navy appear to have considered these balance issues the most.

Finally, all these organizations see the need for certification or validation of personnel, with all the navies and to some extent the U.S. Coast Guard applying a graduated process of underway validation to engineering watches and ships' teams as a whole.

List of PQS Line Items Satisfied by PEO Gas Turbine Class of Instruction (COI) (A-4H-0064) at Surface Warfare Officer School

Upon satisfactory completion of the course, students will have demonstrated the knowledge and skills needed in order to fulfill the requirements of NAVEDTRA 43514-0C, DDG-51 class EOOW to include the following sections: 100 Fundamentals & 200 Systems, which fulfill watch station requirements for line items 301.1.1 and 301.1.2. The class of instruction also satisfies EOOW Tasks, Abnormal Conditions, and Emergencies to include the following line items:

Tasks

301.2.42 Conduct visual test on fuel sample

301.2.45 Shift fuel service tanks suction and return valves

301.2.51 On start GTM from SCU

301.2.53 Online start GTM from SCU

301.2.54 Normal stop GTM from SCU

301.2.55 Operate throttles in lockout manual

301.2.59 Transfer throttle control from SCU to BCU and back

301.2.60 Transfer control from SCU to PACC and back

301.2.69 Remove load and secure SSGTG

301.2.70 Parallel oncoming SSGTG to switchboard bus

301.2.71 Start, monitor, and secure SSGTG

301.2.72 Shift from ship's power to shore power

301.2.72 Shift from shore power to ship's power

301.2.74 Split electric plant

301.2.75 Parallel electric plant

301.2.80 Start, monitor, and secure fuel service pump

301.2.81 Start, monitor, and secure lube oil service pump

301.2.82 Start, monitor, and secure electric CRP pumps

301.2.83 On start GTM

301.2.84 Online start GTM

301.2.85 Normal stop GTM

301.2.86 Transfer control from PACC to SCU and back

301.2.87 Transfer control from PACC to BCU and back

301.2.88 Operate prairie and masker air system

301.2.89 Waterwash GTM

301.2.90 Shift F/O service tanks suction and return valves

301.2.91 Start and stop SWS pumps

301.2.92 Conduct motor and fuel purge GTM

301.2.93 Conduct transparency test on lube oil sample

301.2.94 Demonstrate operation of plasma display and keyboard

301.2.95 Modify and transmit alarm table

301.2.96 Update date and time

301.2.97 Monitor plant status

301.2.98 Shift from auxiliary operation to underway

301.2.99 Shift from receiving shore services to auxiliary operation

301.2.100 Shift from receiving shore services to underway

301.2.101 Shift from underway to auxiliary plant operations

301.2.102 Shift from underway to receiving shore services

301.2.103 Shift from underway on one shaft to underway on two shafts

301.2.104 Shift from underway on two shafts to underway on one shaft

301.2.108 Conduct visual sediment test on lube oil sample

301.2.109 Conduct visual test on lube oil sample

301.2.110 Authorize placement and removal of equipment tag-out

301.2.112 Read and interpret operating logs

301.2.114 Start and stop fire pumps from DCC/TAC-4

Abnormal Conditions

301.4.1 False starts

301.4.2 PT speed loss shutdown

301.4.3 High GTM lube oil temperature

301.4.4 GTM fuel filter differential pressure high

301.4.7 GTM lube oil level low

301.4.9 PT speed high

301.4.10 Failure to light-off/flame-out trip

301.4.11 GG/PT vibration high

301.4.14 Speed limiting active

301.4.15 Module fire system supervisory alarm

301.4.16 Fuel pump A/B discharge pressure low

301.4.17 Fuel prefilter differential pressure high

301.4.18 Fuel filter/separator differential pressure high

301.4.19 Fuel purifier malfunction

301.4.20 Fuel header pressure low

301.4.21 CRP hydraulic flow high

301.4.22 CRP sump tank temperature low

301.4.23 CRP sump level low

301.4.24 CRP hydraulic pressure low

301.4.25 CRP hydraulic flow low

301.4.26 CRP oil head tank level low

301.4.27 Brake air supply pressure low

301.4.28 MRG lube oil sump level low

301.4.29 Lube oil heater temperature high

301.4.30 Lube oil settling tank temperature high

301.4.31 Propulsion Turbine Module (GTM) Emergency Cooldown Procedures

Emergencies

301.5.1 Loss of PLA

301.5.2 High T5.4

301.5.3 Excessive propulsion turbine vibration

301.5.4 GTM stall

301.5.5 GTM cooling air system failure

301.5.6 GG/PT overspeed

301.5.7 Loss of fuel oil pressure

301.5.9 Post shutdown fire in GTM

301.5.10 Hot bearing in MRG/thrust bearing

301.5.11 Hot line shaft bearing

301.5.12 Loss of lube oil pressure in MRG

301.5.13 Main lube oil pump failure

301.5.14 Unusual noise or vibration in MRG or shafting

301.5.15 Unusual noise or vibration in SSGTG

301.5.16 SSGTG overload

301.5.17 Loss of SSGTG

301.5.18 High turbine SSGTG inlet temperature

301.5.19 Post shutdown fire in SSGTG

301.5.20 Loss of EPCC control

301.5.21 Loss of CRP control

301.5.24 Class Bravo fire in GTM

301.5.25 Class Bravo fire in SSGTG

301.5.26 Class Charlie fire in generator

301.5.27 Class Charlie fire in switchboard

301.5.30 Class Charlie fire in the electrical distribution system

301.5.31 Loss of L/O pressure to Gas Turbine Generator

301.5.32 Post shutdown fire in Redundant Independent Mechanical Starting System (RIMSS)

301.5.33 Overspeeding Gas Turbine Generator

301.5.34 Major flooding

301.5.35 Major flammable liquid leak

301.5.37 Loss of SCU AN-UYK-44 / Fault in SCU VME bus

301.5.38 Loss of electrical power

Bibliography

Applied Research International, "Engine and Propulsion–Full Mission Engine Room Simulator," undated. As of June 25, 2008:
http://www.ariworld.com/simulation/engine_propulsion_full_mission_1.asp

Bath Iron Works Corporation, "Propulsion Operating Guide For DDG 83 USS Howard," Washington, D.C.: Naval Seas Systems Command, July 20, 2002.

Clark, Summer, "Modernizing the DDG Fleet—A Total Ship Systems Approach," Wavelengths Online, May 17, 2005. As of March 26, 2008:
http://www.dt.navy.mil/wavelengths/archives/000164.html

Commander, Naval Surface Force, "Guidance for Cruiser-Onboard Trainer Pierside Engineering Certifications and Assessments," Naval Message, July 30, 2008.

Department of the Navy, OPNAVINST 9200.3, Op-04P, "Engineering Operational Sequencing System (EOSS)," September 21, 1976.

———, NAVEDTRA 14300, "Navy Instructional Theory," August 1992.

———, CINCPACFLTINST 9261.1, "Ship's Fuel Maintenance System," December 13, 1993.

———, OPNAV Notice 1500.57A, N869, "Surface Warfare Training Strategy," August 3, 1999.

———, COMNAVSURFORINST 3540.2, "Surface Force Engineering Readiness Process," March 8, 2002.

———, COMNAVSURFOR Instruction 3502.1C, "Surface Force Training Manual," January 1, 2006a.

———, COMNAVSURFOR Message 101516Z, "Maintaining Warfare Certifications," May 2006b.

———, OPNAV Notice 3000.15, N09, "Fleet Response Plan (FRP)," August 31, 2006c.

————, COMSECONDFLT 07-001F, "FY07 Fuel Management Guidance 07-001F," November 2006d.

————, COMNAVSURFORINST 3502.1D, "Surface Force Training Manual," Change 1, July 1, 2007.

————, NAVEDTRA 43514-OC, "Personnel Qualification Standard," Naval Education and Training Command, June 2008a.

————, COMNAVSURFORINST 3540.3A, "Engineering Department Organization and Regulations Manual (EDORM)," September 22, 2008b (with change transmittal 1).

Ebbutt, Giles, "Royal Navy's Simulation Aims for More Than a Semblance of Reality," *International Defence Review*, November 1, 2007.

Etnyre, VADM T. T., Commander, Naval Surface Forces, "Message from Commander, Naval Surface Forces," *Surface Warfare,* Vol. 32, No. 2, Spring 2007.

Jane's Information Group, Jane's Simulation and Training Systems database, 2007. As of June 1, 2007:
http://jsts.janes.com/public/jsts/index.shtml

Lundquist, Edward, N86 Public Affairs, "Simulators Offer Challenges to Improve Real Skills," *Surface Warfare*, Vol. 32, No. 2, Spring 2007.

National Research Council, *Simulated Voyages: Using Simulation Technology to Train and License Mariners,* Washington, D.C.: National Academy Press, 1996.

Naval Air Warfare Training Systems Division, Training System Support Document (TSSD), Orlando, Fla., July 1, 2008.

Naval Center for Cost Analysis, "Naval Visibility and Management of Operating and Support Costs (VAMOSC) 5.0: Detailed Ships—User Manual," February 28, 2006.

Naval Service Training Command, "Battle Stations 21," undated. As of July 11, 2008:
http://www.navy.mil/search/display.asp?story_id=30434

Office of the Under Secretary of Defense (Acquisition and Technology), *Modeling and Simulation (M&S) Master Plan,* Washington, D.C., October 1995.

O'Rourke, Ronald, "Navy Aegis Cruiser and Destroyer Modernization: Background and Issues for Congress," Congressional Research Service Report for Congress, RS22595, February 1, 2007.

Salas, Eduardo, and Janis A. Cannon Bowers, "The Science of Training: A Decade of Progress," *Annual Review of Psychology*, Vol. 52, 2001, pp. 471–499.

Schank, John F., Harry J. Thie, Clifford M. Graf, II, Joseph Beel, and Jerry M. Sollinger, *Finding the Right Balance: Simulator and Live Training for Navy Units,* Santa Monica, Calif.: RAND Corporation, MR-1441-NAVY, 2002. As of March 26, 2008:
http://www.rand.org/pubs/monograph_reports/MR1441/

Yardley, Roland J., Harry J. Thie, John F. Schank, Jolene Galegher, and Jessie Riposo, *Use of Simulation for Training in the U.S. Navy Surface Force,* Santa Monica, Calif.: RAND Corporation, MR-1770-NAVY, 2003. As of June 19, 2008:
http://www.rand.org/pubs/monograph_reports/MR1770/

Yardley, Roland J., Harry J. Thie, Christopher Paul, Jerry M. Sollinger, and Alisa Rhee, *An Examination of Options to Reduce Underway Training Days Through the Use of Simulation,* Santa Monica, Calif.: RAND Corporation, MG-765-NAVY, 2008. As of September 16, 2008:
http://www.rand.org/pubs/monographs/MG765/